# Regulation of Endothelial Barrier Function

# Integrated Systems Physiology: from Molecule to Function to Disease

**Editors**

**D. Neil Granger**, *Louisiana State University Health Sciences Center–Shreveport*

**Joey P. Granger**, *University of Mississippi Medical Center*

Physiology is a scientific discipline devoted to understanding the functions of the body. It addresses function at multiple levels, including molecular, cellular, organ, and system. An appreciation of the processes that occur at each level is necessary to understand function in health and the dysfunction associated with disease. Homeostasis and integration are fundamental principles of physiology that account for the relative constancy of organ processes and bodily function even in the face of substantial environmental changes. This constancy results from integrative, cooperative interactions of chemical and electrical signaling processes within and between cells, organs, and systems. This eBook series on the broad field of physiology covers the major organ systems from an integrative perspective that addresses the molecular and cellular processes that contribute to homeostasis. Material on pathophysiology is also included throughout the eBooks. The state-of the-art treatises were produced by leading experts in the field of physiology. Each eBook includes stand-alone information and is intended to be of value to students, scientists, and clinicians in the biomedical sciences. Since physiological concepts are an ever-changing work-in-progress, each contributor will have the opportunity to make periodic updates of the covered material.

**Published titles**

(for future titles please see the Web site, www.morganclaypool.com/page/lifesci)

Regulation of Endothelial Barrier Function
Sarah Y. Yuan and Robert R. Rigor
www.morganclaypool.com

ISBN: 9781615041206  paperback

ISBN: 9781615041213  ebook

DOI: 10.4199/C00025ED1V01Y201101ISP013

A Publication in the Morgan & Claypool Life Sciences series

*INTEGRATED SYSTEMS PHYSIOLOGY: FROM MOLECULE TO FUNCTION TO DISEASE*

Book #13

Series Editors: D. Neil Granger, LSU Health Sciences Center, and Joey P. Granger, University of Mississippi Medical Center

**Series ISSN**
Print  2154-560X  Electronic 2154-5626

# Regulation of Endothelial Barrier Function

**Sarah Y. Yuan and Robert R. Rigor**
University of California Davis

*INTEGRATED SYSTEMS PHYSIOLOGY: FROM MOLECULE TO FUNCTION TO DISEASE #12*

MORGAN & CLAYPOOL LIFE SCIENCES

# ABSTRACT

The vascular endothelium lining the inner surface of blood vessels serves as the first interface for circulating blood components to interact with cells of the vascular wall and surrounding extravascular tissues. In addition to regulating blood delivery and perfusion, a major function of vascular endothelia, especially those in exchange microvessels (capillaries and postcapillary venules), is to provide a semipermeable barrier that controls blood–tissue exchange of fluids, nutrients, and metabolic wastes while preventing pathogens or harmful materials in the circulation from entering into tissues. During host defense against infection or tissue injury, endothelial barrier dysfunction occurs as a consequence as well as cause of inflammatory responses. Plasma leakage disturbs fluid homeostasis and impairs tissue oxygenation, a pathophysiological process contributing to multiple organ dysfunction associated with trauma, infection, metabolic disorder, and other forms of disease. In this book, we provide an updated overview of microvascular endothelial barrier structure and function in health and disease. The discussion is initiated with the basic physiological principles of fluid and solute transport across microvascular endothelium, followed by detailed information on endothelial cell–cell and cell–matrix interactions and the experimental techniques that are employed to measure endothelial permeability. Further discussion focuses on the signaling and molecular mechanisms of endothelial barrier responses to various stimulations or drugs, as well as their relevance to several common clinical conditions. Taken together, this book provides a comprehensive analysis of microvascular endothelial cell and molecular pathophysiology. Such information will assist scientists and clinicians in advanced basic and clinical research for improved health care.

## KEYWORDS

microvascular permeability, endothelial barrier, cell–cell junction, cell–matrix adhesion, cytoskeleton, signal transduction

# Contents

# Acknowledgments

The authors are grateful to Olesya Litovka and Chris Pivetti for figure preparation and assistance in preparing this text.

# CHAPTER 1

# Introduction

The vascular endothelium is a layer of closely connected endothelial cells lining the inner surface of blood vessels. The endothelium functions as a major barrier at the interface between the blood and tissue by limiting entry of plasma, cells, and molecules from the circulation into the organ parenchyma. In the microcirculation, an important function of microvascular endothelium is controlling blood perfusion by regulating vasomotor response to vasoactive hormones or metabolites to ensure adequate nutrient supply that meets the metabolic demand of tissues. The semipermeable property of the endothelium comprising the wall of capillaries and postcapillary venules enables plasma fluid, nutrients, and cells to move into surrounding tissues and metabolic byproducts to be taken into the circulation. These microvessels serve as the major site for blood–tissue exchange and thus are crucial in fluid and metabolic homeostasis. Appropriate regulation of microvascular fluid hydrodynamics and endothelial barrier function is vital to support normal tissue viability and organ function.

Dysregulation of endothelial barrier function occurs in a variety of disease states and injurious conditions, including inflammation, trauma, ischemia/reperfusion injury, diabetes mellitus, epilepsy, multiple sclerosis, thrombosis, metastatic tumor development, infectious disease, and sepsis, as well as during exposure to certain drugs or toxicants. Endothelial barrier dysfunction is characterized by leakage of fluid, proteins, or small molecules, measured as excessive flux of these molecules across the endothelium (termed *hyperpermeability*), and clinically manifests as accumulation of plasma-like, protein-rich fluid in the extravascular space leading to tissue swelling (termed *edema*). Such a leak response is often accompanied by blood cell (leukocyte) transendothelial/transvascular migration (termed *diapedesis*) and subsequent infiltration into the surrounding tissues. Excessive loss of blood fluid causes reduced blood volume and hence inadequate tissue perfusion. In worse cases, such as severe burn injury, hypovolemic shock occurs. Fluid accumulated in the interstitium compresses tissues and impairs blood–tissue exchange. During this process, a large group of inflammatory mediators are produced or elaborated, including reactive oxygen species (ROS), proteolytically or catalytically active enzymes, amines, and lipid metabolites, rendering an exaggerated host defense response that attacks normal cells and tissues. The pathological consequence of these otherwise bona fide events is tissue injury and organ failure. Although endothelial barrier dysfunction is a common and key process underlying many diseases, it is exceptionally difficult to treat clinically owing to the lack of effective therapeutic interventions. Successful development of target-oriented strategies has been hampered by the many obstacles in translating bench science to bedside care, and by our incomplete understanding of the precise mechanisms that regulate endothelial barriers.

Over the past decades, extensive research has been directed toward cell signaling events and molecular reactions that regulate endothelial barrier structure and function at different levels ranging from cultured endothelial cells to intact isolated microvessels and live animals. Together with clues from human studies and clinical tissue samples, many new theories have been established regarding the cellular and molecular pathways leading to endothelial barrier regulation in health and disease. In regard to the mechanisms altering endothelial barrier integrity in disease states or inflammatory conditions, it is now recognized that multiple cellular pathways and redundant molecular events, with tremendous crosstalk and feedback control, contribute to increased endothelial permeability. Thus, interpretation of its regulatory mechanisms must account for multiple factors, such as the complexity of signaling in real biological systems, cell/tissue heterogeneity, and disease-specific pathophysiology.

The goal of this book is to provide an overview of microvascular endothelial barrier structure and functions. The basic physiological principles (e.g., Starling forces) that govern fluid and solute exchange in the microvasculature will be briefly discussed, followed by a detailed analysis of endothelial barrier structures (cell–cell and cell–matrix interactions). Experimental models that have been employed to investigate endothelial barrier properties will be described. Then the discussion will focus on the signal transduction events in endothelial permeability responses and their relevance to clinical conditions. This information can assist in our further understanding of endothelial molecular pathophysiology for effective treatment or prevention of clinical problems stemming from endothelial barrier dysfunction.

· · · ·

CHAPTER 2

# Structure and Function of Exchange Microvessels

## MICROVASCULAR BLOOD–TISSUE EXCHANGE

The circulatory system delivers oxygen and nutrients to tissues and removes $CO_2$ and other metabolic wastes from tissues, a process conducted at two levels: the macrovasculature and microvasculature [262]. The macrovasculature is composed of arteries and veins, large capacity vessels responsible for transporting blood rapidly toward or away from organs. The microvasculature consists of three types of small vessels: arterioles, capillaries, and venules. These microvessels form a network that regulates local blood perfusion and conducts blood–tissue exchange [262, 415] (Figure 1).

Arterioles are resistance microvessels enveloped by vascular smooth muscle that via contraction or relaxation controls the vessel caliber and thus the volume of blood flow. When arterioles dilate, downstream blood flow is increased. When arterioles contract, blood flow to the downstream microvascular bed (capillaries) is reduced or may be shunted through metarterioles directly to the venous circulation, bypassing the capillary bed. Humoral factors (such as vasopressin and angiotensin) or metabolic factors (such as tissue pH and nitric oxide) can cause vasoconstriction or vasodilation thereby controlling the level of perfusion to meet the metabolic demand of tissues.

While arterioles are a key determinant of local blood flow, which controls the volume for exchange, the exchange process mainly occurs downstream of arterioles in capillaries and postcapillary venules [262]. The walls of these microvessels are thin, mainly composed of endothelial cells and lack vasomotor function due to the absence of continuous smooth muscle. Although other types of cells, such as pericytes, fibroblasts, and smooth muscle cells, are found in the outer wall of capillaries and postcapillary venules, they vary in composition, extent, and function depending upon anatomical location within the microvasculature, as well as organ or tissue type. In the microvasculature of most organs, including the lungs, heart, skeletal muscle, and gut, the walls of capillaries are formed by a continuous (without fenestration) layer of endothelial cells that closely connect to each other, permitting water and solutes <3 nm in molecular radius to pass through freely and restricting the passage of larger molecules. Molecules of >3 nm can move across the endothelium partially and selectively [262]. The relatively high basal permeability and the large surface area of these microvessels provide an efficient means for blood–tissue exchange.

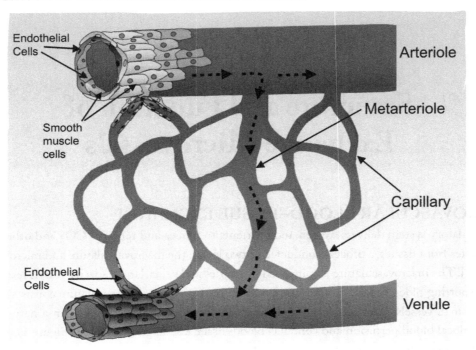

**FIGURE 1:** Architecture of a microvascular bed. Blood flow to capillaries is supplied by arterioles, resistance vessels that are surrounded by vascular smooth muscle. Blood entering the capillary bed is controlled by contraction or dilation of arterioles and precapillary sphincters. Blood flows into the capillaries and is carried out to the venous circulation by postcapillary venules. Alternatively, blood may be shunted directly from the arterioles to the venules via metarterioles, bypassing the capillary bed. These mechanisms control the hydrodynamic pressure and volume of blood flow through individual capillary beds to meet the metabolic and respiratory demands of the tissues.

In short, the conventional concept regarding fluid and solute exchange in the microcirculation is that arterioles do not participate in the exchange process. Only capillaries and postcapillary venules are considered to be exchange microvessels, where capillaries serve as the major site for fluid passage, and postcapillary venules are the primary location for leukocyte diapedesis and plasma protein leakage, processes often seen under stimulated or inflammatory conditions.

## PHYSICAL FORCES GOVERNING FLUID FILTRATION

In 1896, Ernest H. Starling stated [428], "whereas capillary pressure determines transudation, the osmotic pressure of the proteids of the serum determines absorption. Moreover, if we leave the frictional resistance of the capillary wall to the passage of fluid through it out of account, the osmotic attraction of the serum for the extravascular fluid will be proportional to the force expended in the production of this latter, so that, at any given time, there must be a balance between the hydrostatic pressure of the blood in the capillaries and the osmotic attraction of the blood for the surrounding fluids."

Starling proposed that fluid movement across the capillary wall is driven by the difference in the hydrostatic pressure ($\Delta P$) generated by the circulating blood fluid and the colloid osmotic pressure ($\Delta \Pi$) exerted by plasma proteins within the vascular lumen relative to that of the interstitial (extravascular) compartment (Figure 2). This theory was verified and further established by Eugene Landis in 1932 [251, 252], leading to the classic Starling–Landis equation:

$$J_v/A = L_p \left[ (P_c - P_i) - \sigma (\Pi_p - \Pi_i) \right], \qquad (1)$$

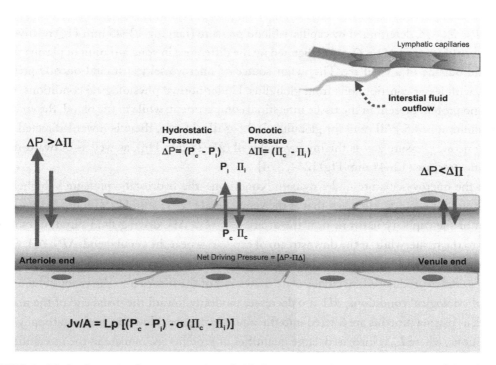

**FIGURE 2:** Hydrodynamic forces governing fluid flow across the microvessel wall (The Starling–Landis model). Fluid movement across capillary walls (flux per unit area; $J_v/A$) is driven by Starling forces: the difference in hydrostatic pressure ($\Delta P$) and colloid osmotic pressure ($\Delta \Pi$), between the capillary lumen (c) and the interstitial compartment (i). Fluid and proteins filtered out of the capillaries accumulates in the interstitial space, decreasing both $P_c$ and $\Pi_c$ and increasing interstitial pressure. Interstitial pressure is decreased by lymphatic outflow. $L_p$ is the hydraulic conductivity and $\sigma$ is the hydrostatic pressure solute reflection coefficient, intrinsic properties of the vessel wall, characterizing sieving or resistance to the passage of solutes (fluid resistance in the case of $L_p$).

where $J_v$ is the fluid volume filtration rate (mL/s), $A$ is endothelial surface area (cm$^2$), $P_c$ is blood fluid hydrostatic pressure inside the capillary (mm Hg), $P_i$ is hydrostatic pressure in the interstitium outside the capillary (mm Hg), $\Pi_p$ is colloid osmotic (oncotic) pressure of the plasma (mm Hg), and $\Pi_i$ is interstitial oncotic pressure (mm Hg). In addition, $L_p$ is the hydraulic conductivity, a coefficient describing the permeation property of capillary wall to water (cm/s/mm Hg), and $\sigma$ is the reflection coefficient, which describes the molecular sieving property of the capillary wall ($\sigma = 0$ means completely permeable without reflection back; $\sigma = 1$ means completely impermeable with 100% reflection back). For example, the reflection coefficient of albumin is ~0.7, meaning that it can cross the capillary wall but its passage is highly restricted. The Starling principle is commonly summarized as:

$$J_v/A = L_p \left[ \Delta P - \sigma \Delta \Pi \right] \qquad (2)$$

where $\Delta P = P_c - P_i$, determined by capillary blood pressure (ranging 12–45 mm Hg) relative to the tissue interstitium, and $\Delta \Pi = \Pi_p - \Pi_i$, dictated by the difference in concentration of plasma proteins inside and outside of a capillary. The major source of microvessel transmural oncotic pressure is albumin, with lesser contributions from globulins. Under normal physiological conditions, there is virtually no protein present in the tissue interstitial compartment, while in the blood, the concentration of albumin is 4.5 g/dL and for globulins is 2.5 g/dL. Hence, there is inward-directed microvascular oncotic pressure across the microvessel wall (25–28 mm Hg), as well as outward-directed hydrostatic pressure (12–45 mm Hg) [264, 374].

In the microvasculature under dynamic conditions, the hydrostatic pressure is higher at the arteriolar end (35–45 mm Hg) than that at the venular end (12–15 mm Hg) (Figure 2) [264]. Thus, upstream of the capillary network near the arterioles, $\Delta P > \Delta \Pi$, driving fluid filtration out of the vessels into the tissue, while at the downstream of capillaries near the venular end, $\Delta P < \Delta \Pi$, favoring absorption of fluid back into the vessel lumen (Figure 2). In this way, vital nutrients are filtered into the tissues, and byproducts of metabolism are drawn out of the tissues into the venous blood. Under normal physiological conditions, $\Delta \Pi$ also decreases modestly toward the distal end of the microvascular bed, as plasma proteins are filtered into the interstitial space. However, under pathophysiological conditions, where $L_p$ is increased, large quantities of proteins accumulate in the interstitial space producing an osmotic/oncotic "sucking" force that drives fluid flow into the tissue and prevents fluid absorption back into the circulation (edema).

Recently, the conventional filtration–absorption model has been challenged by many experiments indicating the absence or transient nature of reabsorption. In particular, Levick and Michel [34] demonstrated that tissue fluid balance could not be maintained by downstream reabsorption in skin more than 10 cm below heart level where venous capillary pressure exceeds oncotic pressure. This is supported by the finding that in the cutaneous venous capillaries or venules at heart level, the interstitial hydrostatic pressure is negative ($P_i = -2$ mm Hg) and the interstitial oncotic pressure is relatively high ($\Pi_i$ = 15.7 mm Hg) so that $\Delta P > \Delta\Pi$, and hence according to this model, no reabsorption can occur.

Most recently, it has been proposed that the Starling principle should be revised to account for the glycocalyx as the principle permeability barrier [204, 304]. Based on physiological evidence produced by Adamson et al. [7, 97], Curry, Levick and Michel [96, 264] proposed a modified version of the Starling principle by revising Eq. (1) to the following:

$$J_v/A = L_p \left[ (P_c - P_i) - \sigma (\Pi_p - \Pi_g) \right] \tag{3}$$

Note the major difference between Eq. (1) and Eq. (3) is that $\Pi_i$ is replaced with $\Pi_g$, the oncotic pressure of the thin layer of interstitial fluid residing between endothelial cells (the intercellular cleft) immediately beneath the glycocalyx network that covers the endothelial surface. In this model, the glycocalyx forms a semipermeable barrier that separates the microvessel luminal compartment from the fluid residing within the [albumin-restricted] space between adjacent endothelial cells (Figure 3).

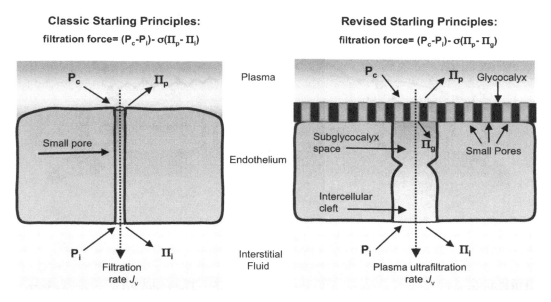

**FIGURE 3:** The revised Starling principle. (Left) In the classic Starling model, fluid flux across permeability pathways is driven by the difference in hydrostatic pressure ($\Delta P$) and colloid osmotic pressure ($\Delta\Pi$), between the capillary lumen (c) and the interstitial compartment (i), as in Figure 2. (Right) In the revised Starling model, fluid flux is driven by the pressure differences between the capillary lumen and the relatively protein-free, volume-restricted space between endothelial cells (intercellular cleft). In this case, the principle barrier to fluid permeability is the glycocalyx network that restricts fluid entry to the intercellular space.

Thus, $\Delta\Pi$ now represents the oncotic pressure difference across the glycocalyx, rather than across the endothelium.

The glycocalyx is a network of glycoproteins and polysaccharides that protects the luminal surface of microvascular endothelium (Figure 4) [38, 438, 461]. The thickness of the endothelial glycocalyx varies by tissue type, from 20 nm to 3000 nm. The glycocalyx is heavily negatively charged and functions to repel blood cells and selectively attracts or repels plasma components according to electrostatic charge. The major components of the endothelial glycocalyx are proteoglycans and proteins with extensive branches of glycosaminoglycans (GAGs) that form the essential structure of the glycocalyx (Figure 4) [264, 301]. Proteoglycans include syndecans with transmembrane pro-

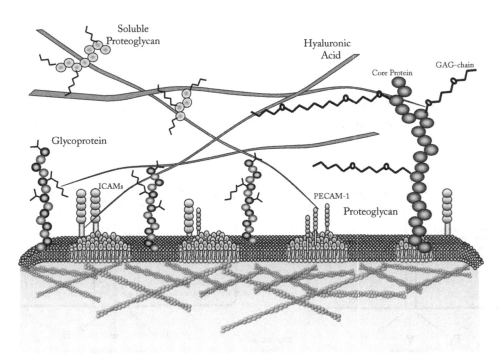

FIGURE 4: Structural components of the endothelial glycocalyx. The endothelial glycocalyx is an extensive network of proteins and polysaccharide polymers that cover the apical (luminal) surface of microvascular endothelium and contribute to barrier function. The backbone components of the glycocalyx are proteoglycans and integral membrane glycoproteins (e.g., adhesion molecules: ICAM-1, PECAM-1) that anchor other glycocalyx components to the endothelium. Hyaluronic acid polymers are woven in amongst the network of glycosaminoglycan (GAG) sidechains attached to proteoglycans. Embedded in this matrix are soluble proteoglycans and other soluble components of the blood plasma.

tein domains and glypicans with glycosylphosphatidylinositol groups that anchor the glycocalyx to endothelium, as well as soluble proteoglycans (perlecan, versican, decorin, biglycan, and mimecan) secreted by the endothelium and embedded within the glycocalyx network [376, 438]. Each proteoglycan core protein may be linked to one or more GAG sidechain composed of heparin sulfate, chondroitin sulfate, dermatan sulfate, or keratan sulfate chains. All are linear polymers of disaccharide subunits: sulfated or unsulfated glucuronic or galacturonic acid, linked to $N$-acetyl-glucosamine or $N$-acetyl-galactosamine. The majority of GAG sidechains for proteoglycans in the endothelial glycocalyx are heparin sulfate, which is 4 times more abundant than chondroitin sulfate. Hyaluronin (hyaluronic acid (HA)) is another type GAG chain polymer that is integral to the endothelial glycocalyx, but is not directly attached to a proteoglycan. HA chains may be either unattached or attached to membrane proteins on the surface of endothelium, contributing to the polysaccharide mesh of the glycocalyx. Integral membrane glycoproteins, including adhesion molecules (selectins, integrins, intercellular adhesion molecules (ICAMs), or platelet/endothelial adhesion molecule 1 (PECAM-1)) provide further structural support for attachment of the glycocalyx the endothelium. Surface expression of these adhesion molecules is dynamically regulated, altering the glycocalyx structure by providing more or less points of attachment to the endothelium and modifying endothelial barrier function. In addition, various soluble proteins secreted by the endothelium or deposited from the blood serum (e.g., albumin) become embedded in the glycocalyx. In some cases, protein deposition can alter microvascular permeability by modifying the charge composition of the glycocalyx. Albumin and other proteins binding further impedes transmural fluid flow by effectively increasing the transmural oncotic pressure above that predicted by the Starling equation, thereby reducing capillary fluid filtration. Consistent with this description, removal of the glycocalyx by enzymatic cleavage increases $L_{\mathrm{p}}$.

The relative presence and importance of the glycocalyx for microvascular fluid exchange varies according to tissue type [264, 301]. It is speculated that glycocalyx plays a particularly important role in restricting transmural fluid flow in discontinuous or fenestrated microvasculature, as in the liver or kidney. Extensively branching glycocalyx can extend into gaps and fenestrations, and thereby limit fluid flow through these openings in the microvessel wall. Curry and Adamson have also shown, through three-dimensional modeling of the endothelial cell–cell junctional ultrastructure in rat mesenteric microvessels, that the 100–300 μm pores that occur periodically between junction proteins create permeability pathways for fluid and solute flux that are too large to account for the apparent reflection coefficient of albumin ($\sigma \approx 0.9$) [97]. Therefore, these investigators hypothesize that the glycocalyx is the principle permeability barrier to proteins across the endothelium in most vascular beds. The glycocalyx may also serve as a sensor of fluid flow in microvascular endothelium [301]. The glycocalyx is modified by shear stress and triggers production of nitric oxide, a potent intracellular signaling molecule that modifies endothelial permeability and barrier function.

# SOLUTE TRANSPORT ACROSS THE MICROVASCULAR WALL
## Convection vs. Diffusion

Thus far, we have discussed the relevant hydrodynamic forces for fluid flux across the microvessel wall during filtration. For solutes (e.g., plasma proteins, salts), two major physical forces drive their movement across a semipermeable membrane: convective flow and solute diffusion (discussed below). For any given set of conditions, either one or the other mechanism will prevail. In most cases in the microvasculature, fluid movement (flux) occurs as a convective flow, the movement of a volume of fluid as a cohesive unit in which bulk fluid flow is driven by hydrostatic pressure, as described by the Starling–Landis relationship [Eq. (1)]. The hydrostatic pressure gradient ($\Delta P$) is opposed by the oncotic pressure gradient ($\Delta \Pi$), which decreases fluid flow ($J_v$) via filtration.

As bulk fluid flow proceeds, plasma proteins (and other solutes) are dragged along in the moving current, a phenomenon called solvent drag [22]. Bulk flow of fluid through a pore is capable of moving dissolved solute particles, irrespective of the solute concentration gradient that otherwise drives solute flux. Hence, any dissolved particle present in the blood may be moved across the endothelium by solvent drag, provided that the size of the permeability pathway is sufficient to permit passage of the solute, determined by the solute reflection coefficient ($\sigma$). As fluid filtration proceeds along the length of the microvascular bed, fluid and albumin are exuded into the tissue interstitial compartment. This leads to a condition where $\Delta P$ is increasingly diminished across the length of the microvessel (Figure 2). At the same time, accumulation of plasma proteins in the interstitium via solute drag increases $\Pi_i$ and decreases the oncotic driving force for fluid reabsorption ($-\Delta \Pi$). According to the Starling–Landis model, the conditions present in real biological situations in microvessels dictate that fluid filtration will persist and that reabsorption is not possible. Clearly, this is not the case, in that under normal physiological conditions, organs and tissues are not maintained in a chronically swollen state. In fact, the measured $P_i$ in vivo is typically negative, and yet fluid and solute reabsorption still occurs in the distal microvasculature. The glycocalyx model and the modified Starling–Landis model accounts for the absence of fluid efflux in the distal microvasculature in real tissues, however, other forces are necessary to drive reabsorption of fluid and solutes [97]. This discrepancy was noted by Starling, who indicated that under conditions of decreased capillary fluid pressure, osmotic forces can temporarily drive reabsorption of salts and water into the blood. When fluid flow becomes sufficiently slowed, as occurs in the distal microvessels, then the prevailing flux mechanism becomes solute diffusion, which then drives fluid reabsorption. Pappenheimer [337] suggested that under conditions where the net thermodynamic force driving fluid flow ($\Delta P - \Delta \Pi$) is near 0, there will be insufficient energy for laminar shear forces to overcome fluid viscosity, and hence bulk fluid flow will not occur. Under these conditions, solutes can diffuse passively across the endothelium along their concentration gradients. Diffusion across a semipermeable membrane is described by Fick's law:

$$J_s = D * A * \Delta C / \Delta x, \tag{4}$$

where diffusive flux of a solute ($J_s$) is driven by the solute concentration gradient ($\Delta C$), according to the diffusion coefficient ($D$) and the surface area of the exchange membrane ($A$), across the distance ($\Delta x$) over which the concentration gradient is dissipated. $D$ is determined by the barrier permeability ($P_s$) to a specific solute, and is inversely proportional to the square root of the solute's molecular mass. This relationship is alternatively expressed as

$$J_s = P_s * A * \Delta C \tag{5}$$

Osmotically active particles (e.g., sugars, salts, amino acids, proteins) attract water, and therefore any net flux of these kinds of solutes across the endothelium is accompanied by a proportional flux of water. This is based on the osmotic pressure, which is proportional to the average thermal kinetic energy of solutes in solution [176]. In a two-compartment system of dilute salt solutions (i.e., physiological concentrations) separated by a semipermeable membrane (freely permeable to water, and not to solutes), the osmotic pressure ($P_{osm}$) exerted by the salt particles in each compartment is proportional to the salt concentration and the absolute temperature (Kelvin) (Van't Hoff's law):

$$P_{osm} = C * R * T \tag{6}$$

where $C$ is the salt concentration, $R$ is the ideal gas constant and $T$ is the absolute temperature. This pressure tends to expand the compartment volume, drawing water molecules into the compartment. In the two-compartment model, if unrestricted by the outer compartment walls, water will be drawn into the compartment of greater salt concentration until the thermal kinetic energy of salts is equalized on both sides (i.e., equal salt concentrations). Based on these principles of osmotic pressure, diffusion of an osmolyte across a semipermeable membrane also causes water flux.

Salt and other solute (e.g., sugars, amino acids) diffusion across endothelium occurs through open intercellular pathways (pores), or may be facilitated by transport proteins residing in the endothelial membrane. This model of permeability is based on Fick's law, yet differs from that originally described by Fick where solute permeability includes partitioning into and diffusion across an artificial lipid membrane with no pores or transporters. Kedem and Katchalsky [225] accordingly adapted Fick's law to describe solute movement across a porous membrane:

$$J_d = PS \, \Delta C, \tag{7}$$

where diffusive flux ($J_d$) is proportional to the solute concentration gradient ($\Delta C$) and the permeability coefficient–surface area product (PS). In real microvasculature, solute fluxes ($J_s$) are driven by both solvent drag (convection; $J_c$) and diffusion such that

$$J_s = J_c + J_d. \qquad (8)$$

Therefore, by incorporating solvent drag forces into Starling's law [Equation (1)], solute flux ($J_s$) becomes [225]:

$$J_s = J_v (1 - \sigma) C_s + PS \, \Delta C. \qquad (9)$$

In real microvasculature diffusion-driven solute, fluxes are short lived and will proceed until the concentration of proteins [oncotic pressure] due to reduced interstitial volume is sufficient to oppose the osmotic force of solute diffusion [97]. In most situations, transvascular solute flux is predominantly driven by convective fluid flow.

## The Capillary Pore Theory

The permeable state of the exchange microvascular wall results from the presence of pores in the endothelium that selectively filters plasma components and retains particles that are too large to pass. In 1951, Pappenheimer, Renken and Borrero [338] published the "pore theory" of capillary permeability. This classical pore theory is based on the principles that molecular movement is affected by steric restriction based on sieving and molecular size. In pore theory, permeability pathways are assumed to be long cylindrical pores through the endothelium (Figure 6), with ideal fluid flow behavior as described by Poiseuille's law, where hydraulic conductivity ($K_F$) is a function of fluid viscosity ($\eta$) and pore radius ($r$):

$$K_F = (A/\Delta x)/(r^2/8 \, \eta). \qquad (10)$$

Based on the measured hydraulic conductivity of individual tracer molecules of various size injected into the dog paw or cat leg, Garlick and Renkin [149] inferred that endothelial permeability pathways consist of small (4 nm) vs. large (80 nm) pores (Figure 5). Smaller pores were predicted to occur at approximately 0.2% of the microvascular surface area, while the decreased permeability to albumin and other large tracers indicated that larger pores are far less abundant [377]. Using this classical two-pore model, one can reasonably account for hydraulic permeability observed in most physiological situations. Some investigators have necessarily refined the original 2-pore model, by adjusting the predicted large pore size (25–30 nm) to fit permeability measurements conducted in various tissues [301, 377]. The physical pathways represented by these pores are uncertain. While the existence of small pores is believed to represent permeability pathways between endothelials at cell–cell junctions, the nature of large pores is somewhat disputed. Speculatively, large pores may represent leaks or gaps that occur infrequently (0.003–0.01% total surface area) and with less consistency than small pores.

**Pore Theory**

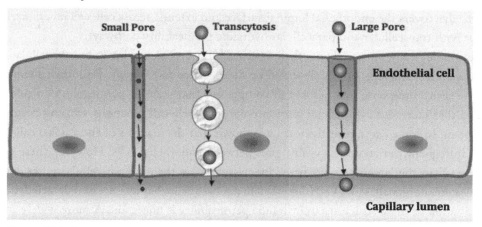

**Poiseuille's Law**
$$K_F = (A/\Delta x)\cdot(r^2/8n)$$

**FIGURE 5:** Pore theory of capillary permeability. (Top) Pore theory predicts the existence of three type of pores that exist in capillary endothelium: small pores represent the normal permeability pathways for fluid and solute flux through intercellular junctions; large pores occur less frequently, and represent permeability pathways for albumin or larger molecules occurring either between endothelial cells (gaps), or across the endothelial cell interior (vesicle transcytosis). Pore theory models permeability pathways as long, ideally cylindrical pores of a precise radius ($r$), length ($\Delta x$), covering a vascular surface area (A), and applies Poisseuille's law for a solution of physiological viscosity ($\eta$) to obtain the filtration coefficient ($K_f$), an indicator of permeability.

Not long after the inception of the two-pore model, ultrastructural studies of endothelium revealed the existence of endothelial vesicles and transendothelial pores that extend across the endothelial cell interior, in addition to physical passages through intercellular junctions [336, 377]. Therefore, large pore phenomena may represent movement of fluid and solutes in intracellular vesicles via transcytosis (Figure 5). Hence, at least three distinct pathways may account for fluid and solute permeability across the microvascular endothelium. Conventional understanding is that transendothelial pores, rather than vesicles, account for most fluid flux across microvessel walls. This is based on the assumption that the effects of solvent drag, i.e., indiscriminant and nonspecific flux of particles proportional to fluid flow, can only be seen in a true transendothelial pore and not in a vesicle-mediated transport system [377].

Modern interpretations of pore theory indicate that the endothelial glycocalyx is responsible for the sieving properties described by pore theory [377]. In this explanation, the glycocalyx serves as a size- and charge-selective filter. The glycocalyx is a meshwork that includes caverns and passages of various sizes, which, taken together, can be modeled as a system of pores of uniform size. The glycocalyx covers the endothelial luminal surface and extends across cell–cell junctions, serving a filter for both transcellular and paracellular hydraulic permeability pathways.

Morphologically, increased endothelial permeability is correlated with several cellular events. Increased vesicle trafficking can be observed by fluorescence microscopy. Endothelial contraction can also be seen in response to a number of compounds that increase permeability. In this case, it is believed that increased cytoskeletal tension pulls apart cell–cell junctions, causing permeability pores to open. In some cases, investigators have shown that the margins of endothelial cells recede, revealing infrequent, yet enormous visible gaps between endothelial cells. However, these gaps are much larger than the large pores predicted by classical pore theory. Transcellular pores, based on vesiculo-vacuolar organelles (VVOs) (discussed in a subsequent chapter), are also detected; however, the contribution of these structures to hydraulic permeability is uncertain. A more widely accepted explanation for hydraulic permeability is pore or gap formation occurring primarily at cell–cell junctions.

# PHYSIOLOGICAL FACTORS AFFECTING FLUID/SOLUTE FLUX

Under physiological conditions, the transvascular movement of fluid and solutes is mainly controlled by three factors: hemodynamics (dictated by the Starling forces), lymphatic drainage (removal of excessive fluid and proteins from tissues), and the barrier (permeability) property of the endothelium.

## Hemodynamics

Microvascular fluid exchange is strongly affected by blood flow. Blood flow in larger microvessels occurs in a laminar fashion and is faster (4.6 mm/s in arterioles; 2.6 mm/s in venules) than in capillaries (0.3 to 1 mm/s) [262, 405]. Flow through capillaries is much more restricted. Because capillaries are smaller in diameter (5–6 μm) than red blood cells (8 μm), red blood cells travel single-file through capillaries, by uniform stacking and bending in the "parachute" configuration [262]. Flow through capillaries is also intermittent, controlled by contraction or vasodilation of precapillary, terminal arterioles. Periodic contraction and vasodilation in microvascular beds (vasomotion) causes the flow through capillaries to increase and subside approximately every 15 sec. Thus, significant periods exist for capillaries where luminal hydrostatic pressure is substantially lower than arterial pressure. These conditions decrease Starling forces, favoring reabsorption. In contrast, increased blood flow through arteriole vasodilation increases capillary luminal pressure and increases filtration. Many signaling molecules and/or byproducts of metabolism increase both arteriole vasodilation and capillary permeability. This has the combined effect of increasing both hydrostatic pressure ($P_c$) and hydraulic conductivity ($L_p$), with both increasing capillary fluid filtration. Therefore, when studying microvessel permeability, one must be careful to distinguish the components of transcapillary fluid flux that are due to increased permeability from those due to arteriole vasodilation and increased blood flow.

## Lymphatic Drainage

Fluid and plasma components that flux across the microvessel wall and are not reabsorbed will accumulate in the tissue interstitial space. In most tissues, lymphatic microvessels are responsible for drainage of excess fluids and solutes (proteins, etc.) that enter the interstitial space (Figure 2) [262]. A few tissues, including bone, cartilage, tendons, ligaments, placenta, and the central nervous system do not have lymphatic vessels [40, 265, 439]. For example, in the brain, excess fluid is removed through the arachnoid space to the ventricles, into the cerebrospinal fluid, and then returned to the blood circulation. In most other tissues, prelymphatic channels [265] shunt excess fluid into initial lymphatic microvessels that are located near the perivascular space of blood microvessels. Initial lymphatic microvessels are formed of juxtaposed endothelial cells surrounded by a discontinuous basement membrane and absence of peripheral contractile cells (i.e., parietal cells or smooth muscle cells) [262]. Lymphatic endothelial cells show minimal expression of cell–cell junction proteins. Rather, lymphatic endothelial cell–cell interfaces extend overlapping flap-like projections that are loosely associated with each other. These flaps are arranged to open in the presence of excess extravascular fluid to allow drainage and to close under normal physiological conditions to prevent backflow. Initial lymphatic microvessels are easily distinguished from blood capillaries because of their irregular nonrounded morphology. Initial lymphatic microvessels form an extensively branched network that feeds into the contractile lymphatic microvessels. Contractile lymphatic vessels (Figure 6) have a rounded morphology and are surrounded by vascular smooth muscle. Lymphatic flow is driven by regular peristaltic contraction of vascular smooth muscle, which pushes fluid through a series of one-way valves within contractile lymphatic vessels. Peristaltic flow returns lymphatic fluid to the lymph nodes, and then through the thoracic ducts into the venous circulation. In skeletal muscle, lymphatic vessels also rely upon skeletal muscle contraction to drive lymphatic flow. In a normal balanced circulatory system, the outflow of fluid via the lymphatic system matches the net influx of fluid into the tissue from the blood circulation and maintains normal homeostatic fluid volume and pressure in the tissue interstitial space. If the lymphatic flow becomes obstructed, or if peristaltic contraction is prevented, excess fluid will accumulate in the interstitial space, a condition known as lymphedema. Lymphedema occurs in certain disease conditions, the most common of which is elephantiasis caused by a parasitic infection that obstructs lymphatic drainage [349].

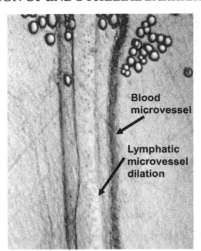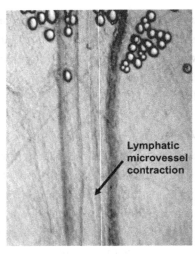

**FIGURE 6:** Lymphatic microvessel contraction. Light microscope images show a contractile lymphatic microvessel undergoing peristaltic dilation (left panel) or contraction (right panel), driving lymphatic flow away from the tissue interstitium and toward the lymph nodes. A blood microvessel is also shown, parallel and adjacent to the lymphatic microvessel. Fatty tissue outside the vessel is seen at the top of each image.

## Endothelial Barrier Properties

The exchange microvessel wall consists of a 1-μm thick monolayer of closely juxtaposed endothelial cells [40, 262]. Adjacent microvascular endothelial cells are joined together at cell–cell (intercellular) junctions, forming the continuous tubular structure of the microvessel lumen. The apical surface of the endothelium exposed to the vessel lumen bears surface glycoproteins (the glycocalyx). The basolateral side of the endothelium is attached to a basement membrane (basal lamina) composed of collagen fibrils, laminin, fibronectin, and glycosaminoglycans. Endothelial cells are anchored to the basement membrane and to the surrounding matrix via cell surface integrins localized at focal adhesions. This structure is semipermeable to water and nonlipophilic molecules and provides size- and charge-selectivity for solute transport across the microvessel wall. The permeability properties of the endothelial barrier are regulated through interactions of endothelial cells, basement membrane and supporting matrix and cells in the surrounding tissue. Barrier dysfunction is often described as increased permeability or hyperpermeability. The consequence of barrier dysfunction is excessive flux of blood fluid, proteins, or cells into the tissue, a pathophysiological process underlying many disease states or injurious conditions.

## Leukocytes and Endothelial Barriers

Endothelial hyperpermeability is a generalized response to inflammation that occurs following trauma, pathogen infection, or chronic disease states [247, 289]. A hallmark of inflammation is extravasation of leukocytes from the blood to the tissue across the microvascular endothelium [71, 267, 330, 383]. Leukocytes are white blood cells circulating in the blood that include lymphocytes (T-cells, B-cells, and natural killer cells), monocytes, and polymorphonuclear granulocytes (neutrophils, eosinophils, and basophils) [268, 317]. Lymphocytes are responsible for the adaptive immune response, which includes production of antibodies against specific antigens, and targeted destruction of pathogens based on antigen recognition [317]. Monocytes and granulocytes, mainly, neutrophils, provide innate immune responses to destroy pathogens, such as bacteria by engulfing them or producing oxygen radicals or enzymes capable of digesting pathogens. Endothelial hyperpermeability, including opening of cell–cell junctions and/or increased vesicle-mediated transcytosis facilitates the movement of leukocytes across the endothelium. In addition, chemical factors secreted by activated blood cells can increase or prolong endothelial hyperpermeability during inflammation [55, 364]. Polymorphonuclear neutrophils (PMNs) present in the blood circulation migrate toward compromised tissues and initiate a generalized immune response [189, 330]. If encountered, PMNs will attack and engulf bacteria into intracellular compartments filled with chemicals (e.g., oxygen-free radicals) and digestive proteases (e.g., metalloproteinases) that dismember and kill the bacteria. PMNs can also migrate across the microvascular endothelium and become activated within the extravascular tissue.

Neutrophil extravasation is a multi-stage process: rolling, activation, adhesion, and transmigration, requiring complex interactions of PMNs or other leukocytes with the microvascular endothelium (Figure 7) [62, 125, 126, 166, 245, 268]. Normal blood flow velocity is extremely fast in microvascular beds ($\geq 1$ mm/sec), preventing sustained interactions of blood cells with the microvessel wall. Under normal physiological conditions, leukocytes in the blood circulation will contact the microvessel wall and interact temporarily through interactions of membrane surface receptors and integral membrane glycoproteins present on both leukocyte and endothelial surfaces. Leukocytes (and platelets) express selectins on the cell surface that bind to sialylated and fucosylated glycoproteins on the endothelial surface, reducing the velocity of leukocytes by approximately 100-fold and causing leukocyte "rolling" along the endothelium [125, 518].

Leukocyte transendothelial migration occurs in response to bacterial invasion or tissue inflammatory injury. In the presence of a compromised microvascular endothelial barrier, leukocytes can become immobilized by firm adhesion to the microvessel luminal surface [255]. This process requires more stable attachment to the endothelium and involves increased expression of adhesion molecules [166, 192], including selectins: endothelial (E)-selectin and platelet (P)-selectin

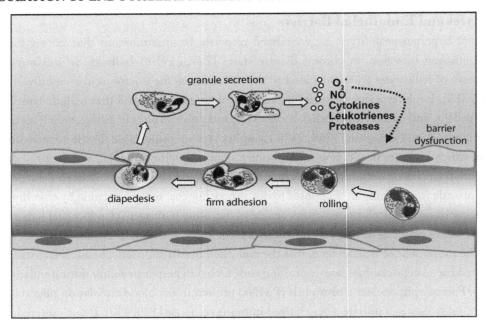

**FIGURE 7:** The stages of neutrophil extravasation. Neutrophil transmigration is a sequential process of (1) rolling along the microvessel wall, (2) firm adhesion to the endothelium (via interactions with cell surface adhesion molecules), (3) diapedesis (transmigration) coordinated by interactions between cell surface glycoproteins, followed by migration into the extravascular space, and (4) neutrophil activation, characterized by granule secretions of hyperpermeability-inducing agents (oxygen-free radicals ($O_2^\bullet$), NO, cytokines, and arachidonic acid metabolites: leukotrienes and prostaglandins), further contributing to barrier dysfunction.

expressed on the surface of endothelium, binding to leukocyte cell surface glycoproteins including P-selectin glycoprotein ligand 1 (PSGL-1) [125, 126, 268]. Firm adhesion is then mediated by binding between leukocyte cell–surface integrins (discussed in a subsequent chapter) and additional cell adhesion molecules on the endothelial surface [125, 126, 166, 192, 268]. Firm adhesion and interactions of endothelial and leukocyte surface adhesion molecules facilitate leukocyte transmigration across the endothelial wall [20].

In general, PMNs are the first leukocyte cell type to arrive at the site of barrier dysfunction [330]. After and/or during transmigration across the microvessel wall, PMNs will become activated and undergo a respiratory burst, characterized by release of granule secretions of numerous compounds [267]. Many of these secretions (e.g., oxygen-free radicals, proteases, nitric oxide, leukotrienes, prostaglandins, cytokines) induce endothelial hyperpermeability [267, 364] and can attack and liquify tissue surrounding the compromised vasculature (pus formation) [330]. PMNs also secrete chemokines or induce endothelial expression of chemokines [307] to attract other leukocytes (macrophages, monocytes, and immune cells) to the site of inflammation. Hence, leukocyte activation and migration across the endothelium are both cause and consequence of endothelial hyperpermeability and barrier dysfunction.

·   ·   ·   ·

CHAPTER 3

# Methods for Measuring Permeability

A variety of methods exist to measure microvascular or endothelial permeability to fluid and/or solutes. Historically, these methods are designed to determine one of several distinctly different, though related mathematical values describing permeability of the capillary wall: the filtration coefficient ($K_f$), hydraulic conductivity ($L_p$), solute permeability coefficient ($P_s$), or the osmotic reflection coefficient ($\sigma$). These parameters relate subtly different properties of solute or hydraulic resistance across endothelial exchange barriers. In many cases, simultaneous measurements of all of the parameters are not feasible, and precise determination of each value is limited by the model system. In particular, quantitative measurements are more challenging to perform in intact physiologically functional tissues. On the other hand, cultured endothelial monolayers may allow greater control of experimental conditions, enabling very refined and quantitative measurements, yet are limited in that they are artificial and inaccurate/inadequate representations of real microvasculature. Between the in vivo and in situ models lies isolated and perfused microvessels, which allow good control of experimental conditions and minimizes interference from nonendothelial cells or systemic humoral factors, enabling precise determinations of permeability in real microvasculature. This chapter presents a discussion of methods commonly used to examine parameters related to fluid and solute permeability of microvascular endothelial barriers.

## ANALYSIS OF FLUID FILTRATION ($K_f$)

Under normal physiological conditions, fluid is filtered out of the blood into the tissues, and removed either by reabsorption, or by outflow to the lymphatic vessels. These processes are maintained in a steady-state that supports normal tissue perfusion and fluid homeostasis. In disease states, excessive plasma fluid and proteins accumulate in the interstitial space (edema), impairing tissue perfusion. Predictions based on Starling's hypothesis indicate that edema can result from imbalances in the hydrostatic and oncotic forces driving fluid flow or from decreased resistance to fluid (increased hydraulic conductivity) across the capillary walls, causing increased filtration. Because of this, early investigators devised methods for estimating the contribution of hydraulic conductivity to tissue edema. These early experiments were based on the assumption that capillary filtration is proportional to Starling forces driving fluid flow ($J_v$) [262]:

$$J_v \propto (\Delta P - \Delta \Pi). \tag{11}$$

However, it was understood that fluid flow must pass through openings in the capillary wall. These openings are a size-selective filter that creates resistance to the passage of fluid and macromolecules. As described in classical pore theory, these resistance pathways can be modeled as cylindrical pores of a characteristic radius ($r$), where fluid flow is dictated by Poiseuille's law [338] described in Eq. (10) such that

$$J_v = A_p\, r^2\, (\Delta P - \Delta \Pi)/(8\, \eta\, \Delta x), \tag{12}$$

where $A_p$ is the total cross-sectional pore area, and $\eta$ is the fluid viscosity, across the length ($\Delta x$) of the permeability pathway. Experimentally,

$$J_v = K_f\, (\Delta P - \Delta \Pi) \tag{13}$$

where the microvascular filtration coefficient ($K_f$) is an empirically determined proportionality constant accounting for all of the physical factors contributing to apparent microvascular transmural fluid resistance, or hydraulic conductivity, within an intact organ or tissue. $K_f$ is proportional to hydraulic conductivity ($L_p$); however, $L_p$ is an intrinsic property of the microvessel wall, normalized to total capillary surface area ($L_p \approx K_f/A$). Therefore, unlike $K_f$, measurements of $L_p$ can be compared between microvessels from different species or from different types of tissues; $K_f$ values must be compared between identical preparations of the same tissue type.

## Measuring $K_f$ in Intact Tissues

The original experiments to measure $K_f$ were performed by Pappenheimer, Renkin, and Borrero [338] in perfused hindlimbs of cats. Amputated hindlimbs of dogs or cats were suspended from a balance to gravimetrically determine changes in tissue fluid content [339]. Both arterial and venous vessels were cannulated and perfused with physiological solutions (filtered blood plasma, concentrated by evaporation, or diluted with Ringer's solution) at controlled fluid pressures (in vs. out) to establish a fixed pressure gradient across the microvascular circulation. Hence both $\Delta P$ and $\Delta \Pi$ could be controlled experimentally. Under these conditions, fluid filtered out of the microvasculature into the tissue was measured as the rate of weight gain of the suspended tissue. These measurements were easily obtained because the hind limb microvasculature is a relatively simple model of endothelial barrier function. Other tissues that include endothelial and epithelial barriers, as well as multiple tissue compartments (e.g., lung or intestine), presented a greater challenge.

Lungs are exceptionally susceptible to edema during local or systemic inflammation or infection. Guyton and Lindsay [174] estimated fluid filtration and $K_f$ in the intact lungs of dogs by controlling the left atrial perfusion pressure, and then measuring differences in wet vs. dry weight of the lung tissue. Differences in wet/dry weights reflected differences in water accumulation in the tissue, proportional to fluid filtration out of the microvasculature. By measuring the arterial and venous blood pressure, as well as colloid osmotic pressure of the blood plasma, these investigators could approximate $K_f$ of the pulmonary microvasculature. However, because fluid accumulation in the lungs in these experiments includes filtration across the endothelium, as well as subsequent filtration across the epithelium into the alveolar air space, these measurements are underestimations of $K_f$. For this reason, Drake, Gaar, and Taylor [114] devised an isolated lung perfusion method similar to that used for hindlimbs. Lungs and heart were removed together (from dogs), and the lungs were mechanically ventilated while the heart was perfused separately through the pulmonary artery and the pulmonary vein. These tissues were perfused by gravity from a reservoir on the arterial side, and outflow was drained to a fluid reservoir on the venous side. Tissues were suspended from a force transducer to measure changes in weight due to fluid accumulation in the tissues. The tissues were initially perfused at a constant pressure until the tissue weight reached steady-state. Next, a small step change in pressure ($\Delta P$) was introduced across the system (by equally elevating both venous and arterial perfusion reservoirs). Following this step change, these investigators observed both a rapid and a slow phase of weight gain (fluid accumulation), which are attributed to microvessel filling and microvascular filtration, respectively. The rate of steady-state weight gain in the slow phase ($\Delta w/\Delta t$) was used to calculate:

$$K_f = (\Delta w/\Delta t)/\Delta P. \tag{14}$$

$K_f$ measurements performed with this protocol were 3- to 4-fold higher than those obtained from wet/dry weights, and, therefore, were considered more likely to represent endothelial filtration and fluid accumulation into the interstitial space. This discrepancy is believed to arise because alveolar epithelial filtration is much more restrictive than endothelial filtration, and the experimental duration is much shorter in the experiments by Drake et al. [114] than in the earlier wet/dry weight experiments [174], suggesting that alveolar filtration had not yet occurred. The type of lung perfusion experiment described by Drake et al. [114] has since been adapted and used to study lungs from other mammalian species, e.g., rabbits [228] or mice [144]. In modern configurations [144], arterial fluids are introduced via a peristaltic pump at a constant flow rate to achieve steady-state tissue weight. Intravascular pressure is increased by switching the height of the venous outflow (via a solenoid valve). Pressure transducers are located at either end of the perfusion circuit (arterial/venous) to confirm that step changes in pressure are equivalent throughout the system. The mean pressure of this system can be obtained by temporarily occluding both ends of the circulation. Under these conditions, in the absence of flow, arterial, and venous pressures will approach an intermediate value representing the mean intraluminal microvascular pressure. The change in this pressure ($\Delta P$) is used to calculate $K_f$, as described by Drake et al. [114].

## ANALYSIS OF HYDRAULIC CONDUCTIVITY ($L_P$)

Eugene Landis originally inferred capillary filtration in vivo by observing circulating red blood cells in intact frog mesentery [251]. He noted that upon physical occlusion of a capillary, red blood cells would continue to migrate in the direction of the obstruction. Hence, fluid continued to flow within the capillary after the occlusion was applied. This could only be explained by the existence of leak pathways across the microvessel wall. This observation by Landis produced the earliest evidence supporting Starling's hypothesis in real microvasculature. With the development of more sophisticated optics and video cameras, Zweifach and Intaglietta [522] applied this method to more precisely measure the fluid filtration coefficient of the microvessel wall:

$$K_f = m/(\Delta P - \Delta\Pi) \qquad (15)$$

where fluid movement ($m$) (i.e., velocity of a red blood cell representing volume flow) is proportional to Starling forces and the filtration coefficient ($K_f$) of the microvessel wall. By injecting macromolecules into the blood circulation, these investigators could adjust the transmural colloid osmotic (oncotic) pressure and demonstrate a linear relationship between Starling forces and fluid flow via filtration.

Based on this observation, Michel, Mason, Curry, and Tooke [306] further developed a modified Landis–Michel micro-occlusion technique for measuring hydraulic conductivity ($L_p$), by cannulating the microvessel with a perfusion pipette, such that hydraulic pressure could be precisely controlled (Figure 8). Using this configuration, the capillary filtration coefficient or hydraulic conductivity ($L_p$) can be determined by measuring flow velocity ($dV/dt$) at different constant perfusion pressures:

$$L_p = (dV/dt)/[A(\Delta P - \Delta\Pi)] \qquad (16)$$

Flow velocity ($dV/dt$) = red cell velocity $\times \pi r^2$, where $r$ is the vessel radius. $A$ is the surface area of the exchange membrane ($\pi \times$ diameter $\times$ length). Because $L_p$ is an intrinsic property of the microvessel wall, normalized per unit surface area, values obtained using this method are directly comparable between microvessels from different tissues.

### Measuring $L_p$ With Colorimetric Dyes

Levick and Michel [263] described an optical method for measuring the fluid filtration coefficient (hydraulic conductivity ($L_p$)) across the wall of an individual perfused microvessel in intact frog mesentery. This method used a colorimetric (densitometric) microscope assay to measure fluxes of

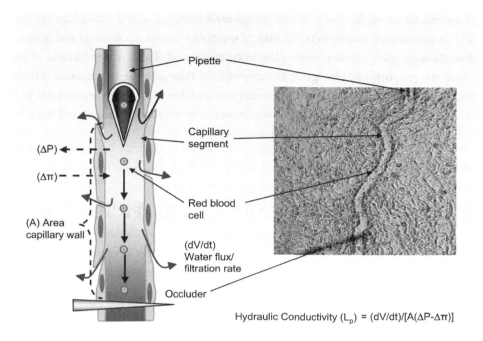

Hydraulic Conductivity ($L_p$) = (dV/dt)/[A($\Delta$P-$\Delta\pi$)]

**FIGURE 8:** The Landis–Michel micro-occlusion technique for measuring hydraulic conductivity ($L_p$) of microvessel walls. An intact tissue mesentery in situ is spread across a microscope field of view to visualize and capture video images of intact microvessels. An "ideal" microvessel segment of uniform diameter, reasonable straightness and absent of side-branches is cannulated at one end and occluded at the opposite end (shown diagrammatically on the left, and as a corresponding image of real tissue on the right). The tissue is bathed in a physiological solution of known oncotic pressure ($\Pi$). Likewise, perfusion solution composition and hydrostatic pressure are formulated to control Starling forces ($\Delta P$–$\Delta\Pi$). Red blood cells (RBCs) are introduced in the perfusate and video recorded to digitally measure fluid flow rate. Following occlusion of the microvessel at the distal end, RBCs will continue to migrate toward the occluder, proportional to the rate of fluid flow *out* of the microvessel through leak pathways in the microvessel wall ($-dV/dt$). Video images are used to assess the physical geometry of the microvessel, to calculate the surface area (A) of the microvessel wall between the cannula and the occluder ($A = \pi\, d\, h$). Changes in $L_p$ are manifested as increased RBC velocity, where $L_p = (-dV/dt)/(A\, (\Delta P - \Delta\Pi))$.

labeled macromolecules across the microvessel wall, under various constant pressure conditions. An individual cannulated microvessel within the mesentery was perfused at one end and occluded at the opposite end. Using this configuration, similar to the red blood cell method, fluid flow into the microvessel can only occur if there is fluid leakage (outflow) across the capillary wall. A colored macromolecule tracer was then introduced into the perfusion solution under constant pressure, and the rate of accumulation of the tracer in the microvessel lumen was determined by optical densitometry. This accumulation rate reflects the flow of solution into the microvessel and is equal to the fluid outflow through leak pathways across the microvessel wall. Upon a step change in perfusion pressure, there is a proportional change in the rate of fluid flow into the microvessel, which can be measured as an increased rate of dye accumulation (optical density) in the microvessel lumen. At two separate fixed, constant transluminal osmotic/oncotic pressures, the measured filtration rates ($J_v/A$) can be used to calculate $L_p$ of the microvessel wall:

$$L_p = (J_v/A)_2 - (J_v/A)_1)/(P_2 - P_1) \qquad (17)$$

# ANALYSIS OF SOLUTE PERMEABILITY COEFFICIENT ($P_S$)

In a real microvascular system, Starling forces and fluid filtration are influenced by the redistribution of solutes (e.g., albumin) across the microvessel wall. Solutes may cross endothelial barriers via diffusion or via solvent drag due to bulk flow of fluids. Both of these mechanisms contribute to the apparent solute permeability ($P_a$) [208, 513], such that

$$P_a = P_d + L_p (1 - \sigma) \Delta P \qquad (18)$$

where $P_d$ is the diffusive component of solute permeability, and $\sigma$ is the solute reflection coefficient.

Huxley, Curry, and Adamson [208] originally described a method for measuring the apparent permeability ($P_a$) of a solute in intact frog mesentery. Microvessels were perfused with a Y-shaped pipette configuration, where one branch of the Y was perfused with a solution containing fluorescent tracer (e.g., TRITC-albumin), and the other with a clear physiological wash solution. Fluorescence in and around the microvessel was monitored with a fluorescence microscope and fluorometer over a window of defined size. Following an equilibration period of constant flow of clear wash solution, the perfusion was switched to solution with fluorescent tracer, and fluorescent intensity ($I_f$) was measured within the optical window. Initially, a step increase in $I_f$ was observed ($\Delta I_f$), corresponding to tracer solution entering the microvessel lumen, followed by a gradual increase ($dI_f/dt$) corresponding to leakage of fluorophore out of the microvessel, across the vessel wall. $P_a$ was then quantified as the ratio of transmural flux of tracer per unit surface area (per unit time), assuming that tracer is proportional to the concentration of solute (albumin):

$$P_a = (1/\Delta I_f) (dI_f/dt)_o (r/2), \qquad (19)$$

where the vessel radius ($r$) is measured directly through the microscope. Next, to solve for diffusive vs. solvent drag components of permeability, $L_p$ was measured by the Landis–Michel micro-occlusion technique (described previously). The hydrostatic pressure ($\Delta P$) in the microvessel, during the flow of fluorophore solution, was determined by adjusting the height of the reservoir with washout solution (with all ends of the Y-connector open) until there was no flow of clear solution from the reservoir in either direction. Then, knowing the osmotic reflection coefficient for the tracer molecule, the components of permeability attributed to diffusive vs. hydraulic pathways could be calculated from equation (18).

## The Isolated, Perfused Microvessel Technique

While the method of Huxley et al. [208] has been very useful for examining microvessel permeability, this procedure is limited in that it can only be applied to semitransparent mesenteric membranes that can be spread across a microscope field of view. For many tissues, this type of preparation is not feasible. Because of the need to study microvascular permeability in parenchymal tissues, Yuan and Granger [513] developed a model measuring albumin permeability in isolated and perfused microvessels (postcapillary venules). Briefly, microvessels of 20 to 70 μm diameter (0.5 to 1.2 mm length) are manually dissected and stripped of extraneous tissue. While submerged in a physiological salt solution, the isolated venule is cannulated with a micropipette-in-pipette at each end of the vessel with each pipette connected to an individual reservoir (Figure 9). One reservoir is filled with physiological wash solution and the other with either the fluorescent tracer (e.g., FITC-albumin) or a chemical treatment. Either reservoir may be selected for perfusion, through a system of appropriate valves. In addition, using the two-cannula system, flow can be arbitrarily directed in either direction as needed. The reservoirs feeding the inflow and outflow pipettes feature adjustable heights such that inflow and outflow pressures can be precisely controlled. Fluorescence video microscopy is used to monitor leaks, and to perform experimental measurements within a selected field of view containing a venule, of known size. Flow rate is determined by an optical Doppler velocimeter by perfusing the system with a dilute suspension of red blood cells. Fixed pressure gradients may be applied across the length of the venule by adjusting the height of the perfusion reservoirs. For example, a pressure gradient ($\Delta P$) of 10 cm $H_2O$ may be achieved by setting the inflow reservoir to 20 cm $H_2O$, and the outflow reservoir to 10 cm $H_2O$. In addition, the oncotic pressure gradient ($\Delta \Pi$) can be controlled with fixed albumin concentration in the perfusion solution. Hence, permeability measurements can be performed under conditions where all relevant Starling forces are known.

The apparent permeability coefficient of albumin is analyzed based on the changes in fluorescence intensity collected from an optical window. Following a step change in fluorescence intensity corresponding to the flow of fluorophore solution through the vessel lumen ($\Delta I_f$), there is a gradual increase in intensity across the entire field of view, corresponding to leakage and the appearance of fluorophore into the bathing solution surrounding the vessel ($dI_f/dt$). The apparent solute permeability coefficient of albumin ($P_a$) is calculated according to Equation (19) (Figure 9) [513]. This method allows for precise measurement of permeability in a real microvessel, yet without

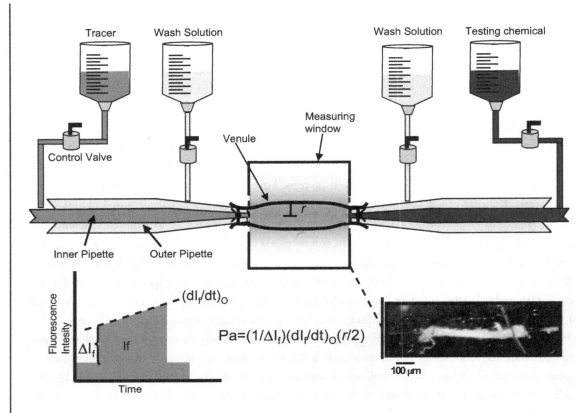

**FIGURE 9:** The isolated, perfused microvessel permeability assay developed by Yuan and Granger. Upper panel: a schematic diagram of the experimental configuration for isolated venule perfusion. Individual venules are manually isolated from intact living tissues (e.g., heart ventricle), and cannulated at both ends with glass micropipettes (real example pictured in lower right quadrant). The living venule is maintained in a physiological bath solution for all manipulations. Using a double perfused "pipette-in-pipette" configuration, perfusions can be rapidly switched, and flow can be introduced in either direction. Typical configurations include separate reservoirs for physiological wash solutions, chemical treatment solution and fluorescent macromolecular tracer solution. Venule luminal hydrostatic pressure is adjusted by raising the height of the perfusion reservoirs. Microscopic analysis is used to measure the venule radius ($r$), and fluorescence intensity within the entire defined field of view ($I_f$) is measured by digital video image analysis. Lower left: a typical experimental protocol includes measurement of baseline (background) fluorescence in absence of tracer, followed by introduction of fluorescent tracer solution. Initially, a step change in fluorescence intensity ($\Delta I_f$) occurs corresponding to fluorescent tracer solution filling the microvessel lumen. This is followed by a more gradual increase in fluorescence intensity ($dI_f/dt$)_o corresponding to leakage of tracer into the bath solution, immediately adjacent to the venule. Venule solute permeability ($P_a$) is calculated: $P_a = (1/\Delta I_f)\,(dI_f/dt)_o\,(r/2)$.

interference from ancillary parenchymal cells or systemic factors that are present in vivo. In addition, the chemical environment (drugs, etc.) and shear conditions (flow velocity, etc.) can be altered as needed, enabling assessments of their direct effects on endothelial permeability.

Yuan and co-workers have subsequently modified this technique for studying leukocyte-endothelium interactions [518]. During inflammation, leukocytes in the blood circulation respond first by binding to the microvascular endothelium, and then crossing the microvessel wall and invading the extravascular tissue (discussed in a later chapter). Using the isolated microvessel perfusion technique, Yuan et al. introduced fluorescent-labeled neutrophils through the inner inflow pipette, at various constant flow rates, and monitored interactions with the microvessel wall using frame-by-frame analysis of fluorescence microscope high-resolution video images. Under normal physiological treatment conditions, neutrophils in contact with the microvessel wall could be seen moving at a slower velocity (rolling), compared to the flow rate of the perfusate (measured with an optical Doppler velocimeter by perfusing red blood cells at 1% hematocrit). Rolling was quantified as the percentage of neutrophils displaying this behavior during the recorded period. Next, these investigators increased the perfusion pressure gradient by equivalently raising and lowering the inflow and outflow reservoirs, respectively, and observed a decrease in rolling as a function of increasing pressure (2.5 to 20 cm $H_2O$). Under inflammatory conditions, activated neutrophils can adhere to and cross the microvessel wall. Following preactivation by exposure to complement factor C5a, neutrophils in the microvessel perfusion assay were seen immobilized next to the microvessel luminal surface (adhesion) in isolated coronary venules. By increasing the perfusion pressure (2.5 to 20 cm $H_2O$), rolling and adhesion events were similarly attenuated in proportion to perfusion pressure. Therefore, adhesion was dependent upon neutrophil activation and normal physiological neutrophil rolling behavior. These investigators also noted that activated neutrophil adhesion occurred in isolated venules, and not in coronary arterioles, supporting the long-held notion that postcapillary venules are the principle site of neutrophil extravasation in the microvasculature. Therefore, the isolated microvessel perfusion method is useful for studying and quantifying leukocyte rolling and adhesion behavior and for observing the differences in these interactions between venules and arterioles.

## ANALYSIS OF THE OSMOTIC REFLECTION COEFFICIENT (σ)

The reflection coefficient accounts for restrictions of solute permeability based on the molecular size of a solute relative to the mean pore size for a permeability pathway. Larger molecules encounter greater resistance and are more frequently "reflected" from the membrane, having a higher value for σ, whereas small molecules are less frequently reflected, and therefore have a smaller value for σ. In most cases, estimates of σ are used based on the diffusion behavior of tracer-labeled solutes. Accurate determinations of σ are variable in real organisms and can be challenging to determine in vivo. An effective method was developed by Laine and Granger [249] in dog hearts, based on the "wash-down" of the lymphatic compartment with greatly elevated perfusion pressure. By elevating vascular perfusion pressure, it was determined that a limit would be reached above which there would be no further decrease in lymphatic protein concentration. This limit occurred when the microvascular permeability pathways to albumin became saturated and a maximal rate of albumin filtration was achieved.

In general, the net solute flux across the microvessel wall ($J_s$) is determined by both diffusive ($J_d$) and convective (solvent drag) ($J_c$) fluxes [249]:

$$J_s = J_d + J_c \qquad (20)$$

This can be re-expressed as:

$$J_s = (1 - \sigma)\, C_p * J_v + PS\,(C_p - C_L) \qquad (21)$$

where $C_p$ and $C_L$ are plasma and lymphatic protein concentration, respectively; $J_v$ is fluid flux; PS represents the protein permeability-surface area product. Solved for $C_L/C_P$:

$$C_L / C_P = [1 - \sigma + (PS/J_v)]/[1 + (PS/J_v)] \qquad (22)$$

At high rates of fluid flow, PS $\ll J_v$, and reduces to

$$C_L/C_P = (1 - \sigma) \qquad (23)$$

Hence, under high-flow conditions, the ratio of lymphatic protein concentration to plasma protein concentration can be used to calculate $\sigma$.

## GENERAL INDICATORS OF PLASMA EXTRAVASATION

In addition to the aforementioned specific permeability parameters, the general status of plasma leakage in vivo is often assessed by tissue staining with colored dyes (e.g., Evans blue) injected into the blood circulation. The dye is allowed time to equilibrate prior to an experimental treatment, following which the animal is sacrificed and dye accumulation is quantified in harvested organs by homogenization and extraction. Typical extraction protocols involve tissue denaturation with trichloroacetic acid or formamide. This procedure was originally described by Miles and Miles [308] using Pontamine Sky Blue and is known as the Miles assay. The Miles assay has been widely used; however, there are several caveats of using this assay [32]. In general, distribution of an intravenous tracer is affected by hemodynamics. For example, treatment with VEGF causes vasodilation, as well as microvascular hyperpermeability [32]. Vasodilation increases the surface area of vessel walls and increases blood flow volume in the microvascular bed. Because most transvascular solute flux is driven by solvent drag and convective fluid flow, vasodilation increases solute flux across the microvessel wall by increasing the volume of fluid in the microvessels, as well as the surface area of the microvascular wall. Capillary recruitment or increased microvascular perfusion density also increase the apparent dye accumulating in the tissue. Distinguishing intravascular from extravascular dye volume requires either an additional indicator of vascular volume or use of a capillary depletion method to determine the fraction of dye that is associated with the vasculature [340].

Evans blue (T-1824) is a 961-Da molecule strongly associated with albumin [19, 30]. Because of this association, Evans blue transport is considered to represent albumin transport. However, the use of Evans blue as a marker for albumin is problematic. The association of Evans blue with albumin is strong in that once binding has occurred, these molecules cannot be readily dissociated [345]; however, the extent of initial binding is highly dependent upon the concentrations of Evans blue and albumin in solution [32]. In many instances, after equilibrating Evans blue with albumin, a substantial fraction of Evans blue remains in solution as free dye. Because of its small size (< 1 kDa), free Evans blue dye permeates through small pores in the microvessel wall much more readily than does Evans blue bound to albumin (>69 kDa). Under these conditions, tissue accumulation of Evans blue will be much higher than that of albumin; therefore, it overestimates the true albumin permeability. In addition, the optical density and absorbance profile of Evans blue is widely variable in plasma from different animal species [19], introducing further inaccuracies to permeability measurements obtained from various animal models. Nevertheless, if thorough binding to albumin is achieved and other potential caveats are addressed, then Evans blue is equivalently effective as a quantitative marker for albumin permeability as radiolabeled albumin for in vitro or in vivo assays [345].

## Intravital Microscopic Measurement of Transvascular Flux

Macromolecules (albumin or dextran) conjugated to a fluorescent probe (e.g., FITC or TRITC) are also frequently used to monitor changes in microvascular leakage in intact tissues conducive to intravital microscopy [183, 328, 487]. Fluorescent-labeled molecules administered intravenously can be visualized in the microvasculature of semitransparent tissues, such as the mesentery, cremaster muscle, and hamster cheek pouch. Ley and Arfors [269] developed a protocol wherein intravenous FITC-dextran was visualized in the microvessels of the hamster cheek pouch by fluorescence microscopy, photographically recorded at various times for densitometric analysis. The greyscale intensities of images of microvessel and extravascular FITC-dextran, as well as a series of cuvettes containing known concentrations of FITC-dextran (standards) were determined by densitometry. These values were then used to calibrate the concentrations of FITC-dextran leakage into the tissues. This method of quantitation was greatly improved by Duran and coworkers [25] who used video recording and digital image analysis to record and calibrate concentrations of FITC-dextran. Using this method, the image optical intensity of a microvessel could be quantified immediately after intravenous injection of FITC-dextran, before any leakage had occurred. Then, knowing the serum FITC-dextran concentration, the concentration of dye leakage outside the microvessel could be determined (provided that fluorescence intensity remained within a predetermined range of linear correspondence to FITC-dextran concentration (0.4 to 3.0 mg/mL in this study). Integrated optical intensities (IOI) are determined by measuring pixel density (and greyscale intensities) of defined regions within an optical window of fixed dimensions and pixel size. IOI values can also be used to express microvascular leakage as the ratio of the transmural intensity difference to the original intraluminal intensity at time = 0 (Figure 10) [205]:

$$IOI_{relative} = (I_i - I_o)/I_i \qquad (24)$$

where regions of interest (ROIs) define intraluminal intensity ($I_i$) or extraluminal intensity ($I_o$) of FITC-dextran, respectively.

Using the same data set, microvascular leakiness can also be expressed as the number of leakage sites per defined area (e.g., 100 $\mu m^2$) in the microscope field of view [205]. Mayhan and co-workers [297] originally described this method using a hamster cheek pouch preparation. FITC-dextran was administered as described in the above preparation, and leaky sites were quantified by acquiring fluorescence microscope images from 10 randomly sampled fields of view within a single preparation at each experimental sampling time. Leaky sites were defined as fluorescent spots greater than 50 $\mu m$ in diameter. Leaky sites counted from 10 fields were expressed as the number of sites per $cm^2$. In this study, the plasma (perfusate) fluorescence and external (superfusate) solution fluorescence were quantified in parallel with imaging, and used to demonstrate disappearance of tracer from the perfusate and appearance of tracer in the superfusate, coinciding with the number of leaky sites. The time dependence and quantitative results from preparations treated with histamine showed excellent correlation between the incidence of leaky sites and the fluorescence leakage measured directly in perfusate/superfusate solutions, indicating the validity of this method.

## ASSESSMENT OF BARRIER FUNCTION IN CULTURED ENDOTHELIAL CELLS

Cultured endothelial cells are often used to model the microvascular endothelium, because of the feasibility of performing complex molecular or pharmacological experiments to study cellular signaling pathways related to barrier function. For example, cultured cells are used to assess solute permeability or electrical resistance across a confluent monolayer of endothelial cells, as indicators of physiological changes in permeability at intercellular junctions. However, these models show

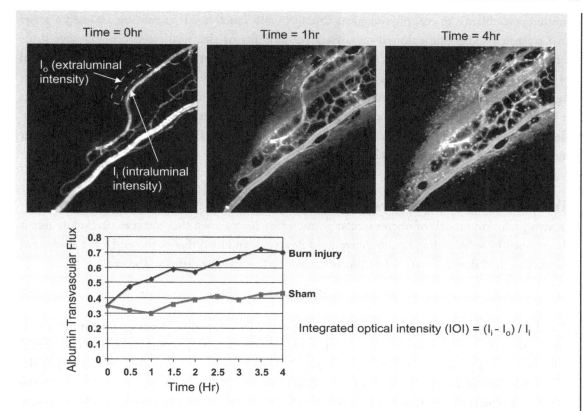

**FIGURE 10:** The use of fluorescence microscopy and digital integrated optical intensity (IOI) to measure macromolecular leakage in intact microvessels in vivo. Upper: fluorescence microscopy images of intact microvessels from mouse gut mesentery with an intravascular fluorescent tracer macromolecule. The appearance of fluorescence in the surrounding tissue can be visualized by fluorescence microscopy and quantified in digital video recorded images, representing microvascular leakage. The intravascular fluorescence intensity ($I_i$) is constant over the experimental period, and can be used to normalize fluorescence intensity due to leakage ($I_o$), such that IOI = $(I_i - I_o)/(I_i)$. Changes in IOI are proportional to changes in transvascular macromolecular solute leakage (flux). Lower graph: in this example, microvessels from mice show increased microvascular leakage (transvascular macromolecule flux) following dermal burn injury compared to anaesthetized, unburned (sham) mice.

limited resemblance to real physiological structures and functions. It cannot be assumed *a priori* that treatments affecting cultured monolayer permeability or electrical resistance will have similar or comparable effects in vivo. Ultimately, all treatments must be tested and validated in real physiological systems. In many cases, cultured endothelial cells are the only viable models for testing mechanistic hypotheses related to cell signaling in control of microvascular barrier function, as it is often impossible or unreasonable to perform these kinds of experiments using in vivo models. Because of this, cultured endothelial cell models are indispensible for fully understanding molecular mechanisms regulating endothelial barriers.

## Transwell Solute Flux Assays

In vivo perfusion models of microvascular permeability have shown that transvascular solute flux is determined by solvent drag in convective fluid flow ($J_v$), and by diffusive forces. In the absence of convective fluid flow ($J_v = 0$), solute flux is determined entirely by diffusive forces:

$$J_d = PS \, \Delta C \tag{25}$$

Based on this relationship, *PS* product can be measured in a zero-flow, 2-compartment tissue culture system where the compartments are separated by a continuous (confluent) layer of endothelial cells grown on a porous filter membrane (Figure 11) [422]. A tracer molecule is introduced to the upper (luminal) compartment at an initial concentration ($C_L$), and the appearance of tracer in the lower (abluminal) compartment is sampled at various time intervals. The time-dependent increase in concentration of tracer in the abluminal compartment ($C_A$) is used to calculate tracer clearance from the luminal compartment, from the slope, expressed as volume flux over time (d$V$/d$t$):

$$V = (C_A * V_A)/C_L \tag{26}$$

where the permeability of the tracer is

$$P_s = (dV/dt)/S \tag{27}$$

or measured over a defined interval (t):

$$P_s = (C_A * V_A)/(t * S * C_L) \tag{28}$$

Malik and coworkers initially used this relationship to determine solute permeability and size selectivity of pores in cultured endothelial cell monolayers [422]. The transwell assay described by these investigators (Figure 11) is widely used to determine transendothelial solute (e.g., albumin, dextrans, sucrose) permeability across cultured endothelial monolayers.

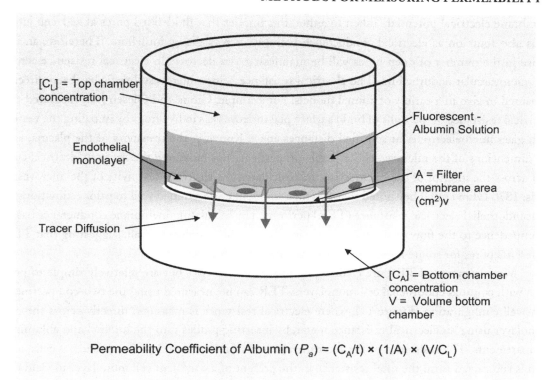

$[C_L]$ = Top chamber concentration

Fluorescent - Albumin Solution

Endothelial monolayer

A = Filter membrane area $(cm^2)$v

Tracer Diffusion

$[C_A]$ = Bottom chamber concentration
V = Volume bottom chamber

Permeability Coefficient of Albumin $(P_a) = (C_A/t) \times (1/A) \times (V/C_L)$

**FIGURE 11:** Transwell solute flux assay for cultured endothelial cell monolayers. Endothelial cells are cultured in growth medium on a porous filter membrane until confluency is attained. This filter membrane forms the floor of a fluid-filled chamber that is then inserted into a microplate well, such that the cell monolayer forms a barrier between upper (luminal) and lower (basolateral) media chambers. During experimentation, a known concentration of tracer-macromolecule is introduced in the upper chamber $(C_L)$, and the increase in tracer concentration is measured in the lower chamber $(C_A)$. These measurements are used to calculate the solute permeability coefficient $(P_a)$ of the monolayer: $P_a = (C_A/t) (1/A) (V/C_L)$, where $A$ is the total surface area of the monolayer, and $V$ is the volume of fluid in the lower chamber.

## Transendothelial Electrical Resistance Measurements

In pore theory, increased transendothelial permeability is due to an increase in the number of open pores at endothelial cell–cell junctions. Open pores are fluid-filled pathways that allow the passage of water and aqueous solutes across the endothelium. In contrast, the endothelial membranes are composed of lipophilic molecules (phospholipids, cholesterol, etc.) that restrict the passage of aqueous solutions, and that function as electrical insulators to maintain the physiological endothelial membrane electrical potential. Taken together, this implies that fluid-filled pores at cell–cell junctions also function as electrical conductance pathways across the endothelium. Therefore, an increase in the number of open pores will be manifested as a decrease in electrical resistance across the microvascular endothelium. The electrical resistance across microvessel walls has been directly measured in vivo in a variety of animal models. For example, Crone and Olesen [92] measured the electrical resistance of the walls of brain surface pial microvessels in live frogs by impaling the vessels with glass microelectrodes at specified distances apart. Knowing the resistivity of the plasma, and the dimensions of the microvessel (i.e., cable properties), they measured the decay in electrical current across the interior of the microvessel, and calculated the electrical resistivity of the microvessel walls: 1870 Ohm cm$^2$. They noted that this measurement corresponded well to prior estimations of transendothelial electrical resistance (TER) performed by calculating membrane conductance based on impedence to the flow of small inorganic ions across the microvessel wall, suggesting that TER represents pores for solute movement across the microvessel wall.

Measurements of TER are challenging to perform in vivo, but are relatively simple to perform with cultured endothelial cell monolayers. TER can be measured using the two-compartment transwell configuration (Figure 11), where electrical resistance is measured directly across the cell monolayer using an electrical resistance meter, by inserting probes into the luminal and abluminal compartments. The resistance of the filter membrane is first measured in the absence of cells, and then is subtracted from the total resistance in the present of a confluent cell monolayer to yield the resistance of the cell monolayer, per unit surface area (specific resistance; Ohm cm$^2$). This method is often used in conjunction with transwell solute flux assays (described above), to determine the monolayer confluency/tightness prior to experimentation, or as an additional measurement to accompany PS determinations.

A more sophisticated, quantitative method for determining the transendothelial electrical resistivity is with an electrical cell–substrate impedance sensor (ECIS) [445]. The ECIS device, invented by Giaever and Keese [156] to quantify cultured cell density and migration, consists of a small cell culture chamber with two gold plate electrodes embedded in the bottom surface (Figure 12). Endothelial cells are grown to confluency in growth medium solution, directly on top of the electrodes. ECIS experiments are performed by applying alternating current (AC) across the electrodes and measuring electrical impedance. Impedance is analogous to resistance, but accounts for all resistances and reactances in an AC circuit. In the ECIS circuit, current flows through the medium and across the cell monolayer, such that the cell membranes function as a capacitor, and the pores or gaps between cells function as resistors. Electrical capacitance is proportional to the amount of cells or confluency of the monolayer. Opening or closing of intercellular gaps or pores across a confluent endothelial monolayer will be manifested as changes in ECIS resistance [445].

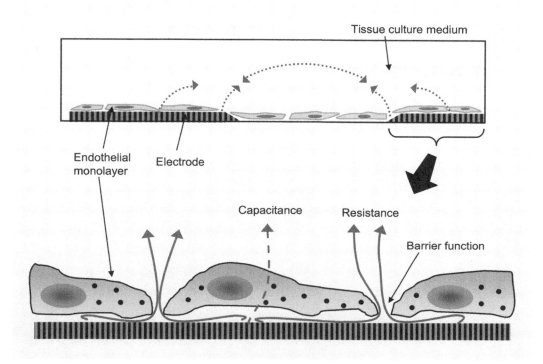

**FIGURE 12:** Electrical cell impedance sensor (ECIS) assay for cultured endothelial cell monolayers. Endothelial cells are grown to confluency on a cell culture plate with two gold electrode contact surfaces. These electrodes are connected to electrical alternating current, and cell impedance is determined. Electrical impedance is equivalent to resistance and can be separated into elements of resistance and capacitance. Capacitance is proportional to the amount of cell material adhering to the electrode, and resistance is inversely proportional to the number of open permeability pathways primarily at cell–cell interfaces or subcellular space due to cell–substrate detachment. The bottom panel is an enlargement of the region of the top panel at the electrode surface.

The resistance component of ECIS measurements includes resistance attributed to fluid-filled spaces between cells ($R_b$) (cell–cell adhesive barrier) as well as that of the fluid volume trapped in the space between the basolateral cell surface and the ECIS electrode ($\alpha$) (indicator of cell–substrate adhesion) [156]. These properties ($R_b$ and $\alpha$) can be determined by examining the impedance at varied frequencies and fitting the data to the mathematical model described by Giaever and Keese. The cell radius ($r$) is an empirically measured value (modeled as a sphere in this example), the medium resistivity ($\rho$) is measured directly in the absence of cells, and the relationship, $\alpha = r * (\rho/h)^{0.5}$ is used to calculate the height ($h$) of the subcellular space. The value of $h$ will increase as cell focal adhesions detach from the electrode surface. Therefore, ECIS can be used to quantitatively model changes in endothelial barrier properties attributed to intercellular gap formation vs. detachment at focal adhesions [223].

CHAPTER 4

# The Endothelial Barrier

The barrier function of exchange microvascular endothelium derives from the integrity of the endothelial structure, which undergoes moment-to-moment changes at the cytoskeleton, cell–cell junction complexes, and cell attachments to extracellular matrix and basement membrane. Appropriate regulation of these events maintains a low and selective permeability to fluid and solutes under normal physiological conditions. Endothelial barrier dysfunction occurs during stimulation by inflammatory agents, pathogens, activated blood cells, or disease states. The pathophysiology is characterized by excessive flux of plasma across the exchange microvessel wall into the surrounding tissues. Traditionally, compromised endothelial cell–cell junctional integrity is considered to account for the leak response. However, recent evidence demonstrates that blood fluid, solutes, and even circulating cells can cross the endothelium via two routes: through the cell body (transcellular), or between the cells (paracellular, or intercellular) (Figure 13). Here, we discuss the ultrastructural basis and function of the transcellular pathway vs. paracellular pathway.

## TRANSCELLULAR PERMEABILITY: VESICULAR TRANSCYTOSIS

Transcytosis represents an important pathway of endothelial transcellular permeability to macromolecules [reviewed in 78, 239, 301, 305, 361]. The mechanism involves vesicle-mediated endocytosis at the endothelial luminal membrane, followed by transcytosis across the cell, and exocytosis at the basolateral membrane. This process can be completed by individual vesicles capable of shuttling from the apical to basolateral membrane of an endothelial cell, or by clusters of interconnected vesiculo-vacuolar organelles (VVOs) that form channel-like structures 80–200 nm in diameter, spanning the cell interior (Figure 13) [119, 237].

Vesicle-mediated transcytosis occurs when albumin binds to gp60 receptors on the endothelial cell surface [201, 312, 469]. Endocytosis and exocytosis have been visualized by electron microscopy in capillaries and postcapillary venules using tracer-labeled (e.g., gold-labeled) albumin [377], and other macromolecular markers [119]. EM micrographs reveal albumin apparently in various phases of transcytosis in endothelial cells: open and closed vesicles at the luminal and abluminal surfaces. Blood cells (leukocytes) may also be enveloped by endocytic vesicles and moved across the endothelial cell interior by transcytosis. This has been shown in microscope images of

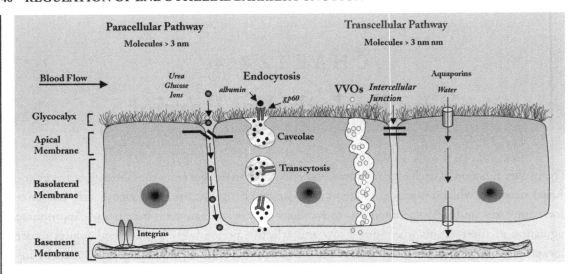

**FIGURE 13:** Transcellular and paracellular permeability pathways across the microvascular endothelium. Barrier function of the microvasculature is provided by closely apposed endothelial cells of the microvessel walls. The thin layer of endothelium is attached to the microvascular basement membrane via endothelial membrane-bound integrins. Endothelial cells are joined together by intercellular junction proteins that allow the selective passage of solutes and fluids across the endothelium. Intercellular junctions can become more porous, or even form large-sized gaps under pathophysiological conditions. The glycocalyx forms a selective filter across the endothelial luminal surface, forming an additional permeability barrier. Solutes can also traverse the cell interior via receptor-mediated vesicle endocytosis originating at caveolae, or via vacuole-vesicular organelles (VVOs) that can fuse with trafficking vesicles and form open transcellular pores. Transcellular water transport can occur in parallel with other fluxes, through regulated water channels (aquaporins) in the endothelial cell membrane.

fluorescent-labeled leukocytes extending foot processes into membrane invaginations, and being internalized [68, 69].

Endocytosis and exocytosis are mediated by caveolae, lipid raft microdomains that form "cave-like" invaginations in the plasma membrane (Figures 13 and 14) [360, 361, 424]. The volume of fluid residing within caveolae constitutes approximately 15–20% that of the endothelial cell interior volume [140], and therefore by forming vesicles, caveolae are capable of moving substantial amounts of fluid and solutes across the cell interior [360]. Caveolae contain caveolin-1 (cav-1), a 22-kDa structural protein that is recruited to the membrane and forms an oligomeric assembly that is necessary for the characteristic shape and structure of caveolae [361, 397].

Caveolae contain typical lipid raft components including cholesterol and sphingolipids, as well as scaffolding and signaling molecules that modify cav-1 function and initiate vesicle traffick-

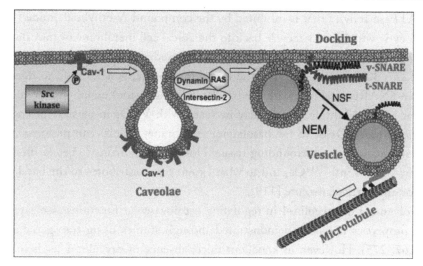

**FIGURE 14:** Vesicle endocytosis and transcytosis across endothelial cells. Vesicle formation is triggered by Src kinase-mediated phosphorylation of caveolin-1 (cav-1) at the endothelial luminal membrane. Cav-1 subunits aggregate in lipid rafts and oligomerize to form caveolae. Other events, including albumin binding to the gp60 albumin receptor trigger vesicle endocytosis. Accessory proteins including dynamin and intersectin-2 are recruited to form the elongated neck of caveolae invaginations, which are then pinched off in response to Ras signaling to form enclosed vesicles. Vesicles remain docked to the inner surface of the cell membrane by vesicular (v)-SNARE binding to membrane-bound target (t)-SNARE, until docking is disrupted by the NEM-sensitive factor (NSF). Undocked vesicles may attach to microtubules via ATP-driven motor molecules (kinesin/dynein) that facilitate vesicle movement across the cell interior. Trafficking vesicles may fuse with VVOs, or may dock (via SNAREs) at the basolateral membrane and undergo membrane fusion and exocytosis, releasing contents into the extravascular space. This entire process is transcytosis.

ing (Figure 14) [78, 362, 424]. Cholesterol is necessary for cav-1 recruitment to the membrane. Oligomerization of cav-1 and caveolae formation are then initiated by phosphorylation of cav-1 (by Src kinase) [260, 271, 311, 408]. Subsequent recruitment of dynamin (large GTPase) and intersectin (adapter protein) allow caveolae to fully form and to pinch off from the plasma membrane [363] forming endocytic vesicles approximately 70 nm in diameter (Figure 14). These vesicles remain docked to the plasma membrane through interactions of soluble $N$-ethylmaleimide-sensitive factor attachment receptors (SNAREs): vesicle-associated (v)-SNARE (vesicle-associated membrane protein (VAMP)) binds to membrane-associated target (t)-SNARE (25-kDa synaptosome-associated protein (SNAP-25) and syntaxin) [200, 362]. Dissociation of v-SNARE from t-SNARE and vesicle detachment are induced by $N$-ethylmaleimide-sensitive factor (NSF), a vesicle fusion protein with ATPase activity that is inhibited by the compound N-ethylmaleimide (NEM) [359]. Detached endocytic vesicles may recycle back to the apical cell membrane or may move across the cell interior to the basolateral side (transcytosis), likely mediated by interactions with cytoskeletal microtubules [277, 301]. Upon arrival at the target membrane, vesicles bind to docking proteins (v-SNARE to t-SNARE) and fuse with the basolateral (tissue side) membrane or at VVOs within the cell interior [361]. Multiple caveoli may be fused with VVOs in clusters in endothelial cells [372]. Once fused to VVOs or to the basolateral membrane, vesicles can perform exocytosis, releasing their contents into the surrounding tissue. However, it remains to be clarified what solutes are deposited by vesicles into VVOs, and to what extent this contributes to the basal or stimulated permeability for any particular solute [119].

The involvement of caveolin-1 in regulating cardiovascular functions associated with endothelial barrier properties has been demonstrated through studies using transgenic and knockout animals [35, 167, 275]. However, in knockout mice, absence of caveolin-1 leads to a compensatory increase in the paracellular permeability response to hyperpermeability-inducing agents [361]. Thus, while caveolin-1 and transcytosis contribute to regulation of endothelial permeability under normal circumstances, transcytosis is not strictly necessary for hyperpermeability responses because of compensatory hyperpermeability at cell–cell junctions. Vogel and coworkers have also shown that although albumin transcytosis occurs in the intact lungs, transcellular albumin flux does not contribute to fluid filtration [469]. Hence, the contribution of endothelial transcytosis to fluid homeostasis under physiological conditions or plasma leakage under pathophysiological conditions in most systems is not known. In contrast, investigators have reported that transcellular water flux accounts for as much as 50% of hydraulic conductivity in some endothelial systems [489]. An alternative mechanism for facilitating the transcellular passage of fluid is through the transmembrane water channel aquaporin [264]. Aquaporins are integral membrane proteins expressed in endothelial cells that permit the diffusive flux of water across the cell membrane. However in many tissues the contribution of aquaporin channels to $L_p$ is believed to be minor or insignificant (<10%) [378].

## PARACELLULAR PERMEABILITY: CELL–CELL JUNCTIONS

The paracellular pathway is responsible for the majority of leakage of blood fluid and proteins across the microvascular endothelium under pathophysiological conditions. In the microvessels of certain types of tissues or organs, such as the kidney and liver, there are discontinuities or fenestrations between endothelial cells that are sufficiently large to permit the passage of large molecules or proteins [262, 305]. In other organs, most endothelial cell–cell interfaces are fused together by intercellular junctions or pores that selectively allow water, macromolecules, and even blood cells to pass through. The structural and functional integrity of these junctions is a major determinant of paracellular permeability.

Two types of intercellular junctions have been characterized as the cell–cell adhesive barrier structures in the microvascular endothelium: the adherens junction (AJ) and tight junction (TJ) (Figure 15) [239, 301]. AJs are found in most, if not all microvascular beds, and are the most ubiquitous type of endothelial cell–cell junction. AJs are impermeant to albumin (69 kDa; molecular radius 3.6 nm) and other large proteins, and thus the major determinant of endothelial barrier to macromolecules in many organs and tissues [301, 423]. Compared to AJs, TJs are less common in the peripheral microvasculature. TJs are mainly expressed in the microvascular endothelium of some specialized tissues, for example, the blood–brain or blood–retinal barriers [184, 340]. TJs impart additional barrier function, preventing the passage of much smaller molecules (<1 kDa), even restricting the flow of small inorganic ions (e.g., $Na^+$).

The barrier properties of endothelial adherens and tight junctions are due to interactions of integral membrane glycoproteins on neighboring endothelial cells (Figure 15). The extracellular domains of junction proteins displayed on the surfaces of juxtaposed endothelial cells bind to each other and form a seal, restricting the passage of molecules between the cells (paracellular). The tightness of this barrier varies according to the types of junctions that are present in each tissue type, and subject to moment-to-moment changes in response to physical forces or biological signals. Deleterious effects on junction integrity may arise through degradation or dissociation of junction

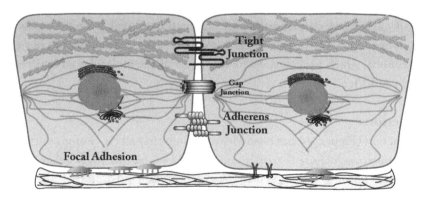

**FIGURE 15:** Endothelial cell junctions and adhesions. Endothelial cells of the microvessel wall are joined together by intercellular junction proteins: adherens junctions (AJs), tight junctions (TJs) and/or gap junctions. Barrier function in most vascular beds is provided by AJs. Some specialized microvascular beds rely upon TJs for additional barrier function. Gap junctions, formed of connexins, primarily facilitate signaling between endothelial cells, and do not directly contribute to barrier function. Endothelial cells are anchored to the basement membrane via focal adhesions. Focal adhesions and intercellular junctions are interconnected via cytoskeleton; Barrier function is dependent upon the stability and integrity of these three elements.

proteins, reorganization or internalization of junction structures, altered interactions with the cytoskeleton, or destabilization of attachments to the ECM that ultimately interfere with the sealing efficiency of intercellular junctions and cause endothelial hyperpermeability. The composition of cell–cell junctions, including the types of junction proteins present, is determined by junction type. In general, the type of junction dictates the pore size and, hence, the size of molecule that is permitted to pass. For example, the mean pore size of adherens junctions is approximately 3 nm [301, 423], whereas the mean pore size of tight junctions is approximately 1 nm [96, 452].

There are two additional structures related to endothelial cell–cell junctions that are not considered to be determinants of paracellular permeability. First is the gap junction (Figure 15). Gap junctions are found mainly in larger vessels and do not contribute significantly or directly to microvascular barriers. The molecular structure of gap junctions is characterized by six units of connexin forming a channel that connects the cytosols of adjacent endothelial cells, allowing rapid propagation of signaling molecules (e.g., $Ca^{2+}$) between the cells [242]. Thus, although gap junctions may indirectly participate in the regulation of endothelial permeability by promoting cell–cell communication, they do not provide the barrier function per se. The second issue concerns fenestrations and discontinuous endothelium. A few tissues contain microvascular endothelium with open fenestrations ranging 50 to 60 nm in diameter (e.g., kidney, intestine or choroid plexus), or discontinuous endothelium with gaps as large as 100 nm (e.g., liver or spleen), permitting the passage of very large solutes [262]. These types of junctions appear to be specialized structures in tissues responsible for absorption of nutrients or detoxification and elimination of toxic wastes. Due to their minimal importance or noncommonality in permeability regulation, these structures are not emphasized in the following sections, which provide a detailed analysis of the molecular structure and functional regulation of AJs and TJs.

## Adherens Junctions

The adherens junction (Figure 16) has been identified in nearly all types of vascular beds, especially in the peripheral microvasculature. Vascular endothelial (VE)–cadherin is believed to be the most important protein in forming the molecular basis, as well as regulating the function of AJs. VE–cadherin is a transmembrane receptor; its extracellular domain binds to the extracellular domain of another VE–cadherin expressed in the membrane of an adjacent endothelial cell. By forming a homotypic bond in this manner, VE–cadherin glues the neighboring cells together. The intermolecular binding of VE–cadherin extracellular domains is dependent upon extracellular calcium. $Ca^{2+}$ binding to negatively charged amino acid residues on the extracellular domain of VE–cadherin promotes the protein conformation necessary for VE–cadherin to perform homotypic binding [386]. Intracellularly, VE–cadherin is connected to the actin cytoskeleton via a family of catenins (α-, β-, γ-, and p120-catenins). A current model of adherens junction structure is that VE–cadherin binds

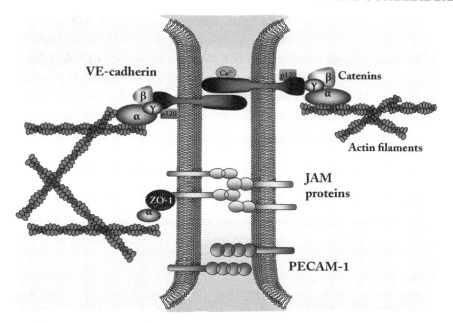

**FIGURE 16:** Adherens junctions. Adherens junctions (AJs) are ubiquitous throughout the vasculature. The intercellular adhesion protein vascular endothelial (VE)-cadherin is principally responsible for barrier function. VE–cadherin homophilic intercellular binding is stabilized by calcium ions and by intracellular connections to the actin cytoskeleton: $\alpha$-, $\beta$-, $\gamma$- and p120-catenin connect VE–cadherin to actin microfilaments. Other junction proteins contribute to AJ structure, including Junction adhesion molecules (JAM-A, -B and -C) and platelet-endothelial cell adhesion molecule (PECAM)-1. JAMs connect to the actin cytoskeleton via zona occludens (ZO)-1 and $\alpha$-catenin, which may stabilize AJs. PECAM-1 facilitates cell–cell binding with circulating blood cells.

directly to $\beta$-catenin and $\gamma$-catenin, which in turn are connected to actin via binding to $\alpha$-catenin [301]. Connection to the actin cytoskeleton is further stabilized by $\alpha$-catenin binding to other proteins, including $\alpha$-actinin, vinculin, vasodilator-stimulated phosphoprotein (VASP) and formin. VE–cadherin is also stabilized by binding to p120-catenin, though p120-catenin does not directly bind to actin. Rather, p120-catenin binds to protein kinases (Src kinases) and phosphatases, serving as a scaffold to bring these signaling molecules into proximity with adherens junction proteins for further interactions [350, 380]. Thus, the catenins not only serve as a structural linkage between VE–cadherin and the cytoskeleton, they also transduce biochemical signals for cell–cell communications. The stability of the VE–cadherin–catenin–cytoskeleton complex is essential to the maintenance of endothelial barrier function [15, 389, 465].

In addition to VE-cadherin and catenins, other proteins are present at cell–cell contacts and may associate or interact with adherens junctions, including E-cadherin, junctional adhesion molecules (JAMs), and platelet–endothelial cell adhesion molecule (PECAM-1). The specific contribution of these proteins to endothelial barrier properties is unclear. JAM proteins can bind to zona occludens 1 (ZO-1), a linker protein that connects tight junction proteins to α-catenin (described below), as well as to signaling molecules and proteins that stabilize the actin cytoskeleton. PECAM-1 binds to integrins on leukocytes and may facilitate leukocyte transmigration across the microvascular endothelium.

## Tight Junctions

Endothelial tight junctions are similar to adherens junctions, but are composed of interactions of tight junction proteins: occludin, claudins (3/5), and JAM-A (Figure 17) [2, 184, 301]. Occludin

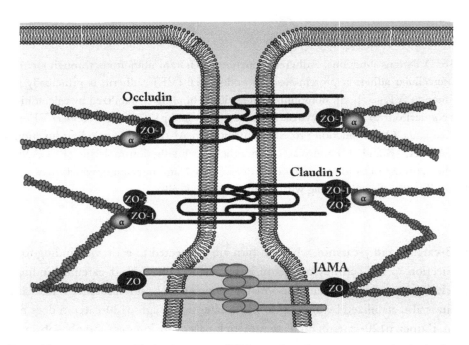

**FIGURE 17:** Tight junctions. Tight junctions (TJs) are found in most vascular beds; however, TJs contribute to microvascular barrier function only in a few specialized tissues, including the brain, retina and testicles. The intercellular junction proteins mainly responsible for TJ barrier function are occludin and claudin-5. TJ protein intercellular binding is stabilized by connections to the actin cytoskeleton via ZO-1, ZO-2 and α-catenin. JAM-A also contributes to TJ structure.

and claudins are integral membrane proteins, each with four transmembrane domains and two extracellular loop domains. The extracellular loop domains of occludin or claudins form homotypic binding with the extracellular domains of like molecules on neighboring endothelial cells. JAM-A, a member of the immunoglobin superfamily of proteins, is also present in tight junctions, though the role of JAM-A in tight junctions is not understood. Occludin, claudins, and JAM-A are connected to the actin cytoskeleton via zona occludens proteins (ZO-1, ZO-2) and α-catenin. In addition to connecting junction proteins to the cytoskeleton, ZO proteins serve as signaling molecules (guanylate kinases) or scaffolding proteins that recruit other signaling molecules via PDZ and Src homology 3 (SH3) binding domains. Hence ZO proteins play both structural and signaling roles in tight junctions. The connection between tight junctions and the actin cytoskeleton is further stabilized by actin cross-linking proteins (e.g., spectrin or filamen) and accessory proteins (e.g., cingulin or AF-6) [184, 414].

## EXTRACELLULAR BARRIERS: FOCAL ADHESIONS

Focal adhesions are points of attachment between the endothelial basolateral membrane and the surrounding extracellular matrix (ECM) of the microvascular wall [493] (Figure 18). The major structural components of focal adhesions are transmembrane receptors called integrins. Integrins are a family of glycoproteins expressed as α/β heterodimers (discussed further in the next chapter). Their intracellular domains interact with the cytoskeleton either directly or indirectly through the linker proteins paxillin, talin, vinculin, or α-actinin (Figure 18, top), and their large extracellular domains bind to respective matrix proteins, such as fibronectin, vitronectin, collagen, fibrinogen, and laminin (Figure 18, bottom) [197, 348]. The molecular organization of integrins varies depending on the chemical and physical states of extracellular matrices [106].

Vascular endothelial cells express multiple integrins with distinct combinations of α/β subunits, including $\alpha_1\beta_1$, $\alpha_1\beta_2$, $\alpha_1\beta_5$, $\alpha_2\beta_1$, $\alpha_3\beta_1$, $\alpha_5\beta_1$, $\alpha_6\beta_1$, $\alpha_6\beta_4$, $\alpha_v\beta_3$, and $\alpha_v\beta_5$. Typically, $\alpha_1\beta_1$ and $\alpha_1\beta_2$ bind to collagen; $\alpha_3\beta_1$, $\alpha_6\beta_1$, and $\alpha_6\beta_4$ bind to laminin; $\alpha_5\beta_1$ binds to fibronectin; and $\alpha_v\beta_3$ and $\alpha_v\beta_5$ bind to vitronectin [14, 287]. Many of these integrins recognize the arg-gly-asp (RGD) sequence in matrix proteins and thereby are able to interact with more than one extracellular ligand.

Integrin–matrix binding is essential to the establishment and stabilization of endothelial barriers [107]. Altering integrin-binding properties reduces focal adhesion strength [448] or causes cell detachment from the substratum [77, 106]. Synthetic peptides that compete the RGD-binding sequence or antibodies directed against the $\beta_1$ subunit of integrins produce a dramatic increase in transendothelial flux of water and large solutes [98, 368, 483]. Direct evidence that underscores the physiological significance of integrin–matrix interactions comes from Wu's study in intact exchange microvessels, showing that inhibition of integrin binding to either fibronectin or vitronectin with

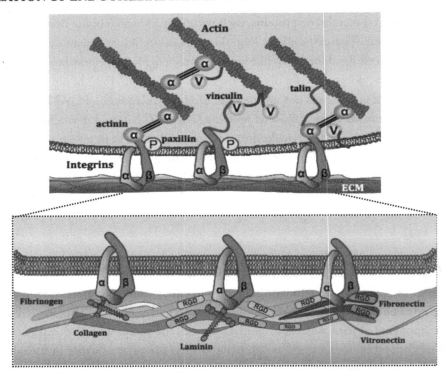

**FIGURE 18:** Focal adhesions. Upper panel: endothelial cells are anchored to the basement membrane via focal adhesions. Focal adhesions are lipid raft microdomains in the endothelial basolateral membrane that are enriched in integrins, transmembrane heterodimeric proteins that bind to the basement membrane. Clusters of integrins are stabilized by intracellular connections to the actin cytoskeleton, via linker proteins (e.g., actinin, vinculin, talin, paxillin). Lower panel: integrins bind to extracellular matrix proteins bearing the amino acid sequence arginine–glycine–aspartate (RGD). RGD proteins (e.g., fibrinogen, fibronectin, vitronectin) are embedded in the network of collagen fibers and laminin that forms the basement membrane. Focal adhesion integrity and binding to the basement membrane prevents endothelial detachment and maintains vascular barrier integrity.

synthetic RGD peptides dose-dependently increased albumin permeability by 2- to 3-fold [495]. The RGD-induced hyperpermeability was time-dependent and reversible upon clearance of the peptides, indicating that the effect was not merely due to a permanent disruption of the endothelium, consistent with the idea that endothelial cell–matrix adhesion is a dynamic process [509, 511].

Integrins provide an important structural support for maintaining endothelial barriers as well as their moment-to-moment changes in coordination with other barrier components. This structural support is multidirectional and may not be limited to the basolateral site of endothelial cells.

Indeed, some members of the integrin family have been identified to be located at endothelial cell–cell borders [250]. It is suggested that these integrins collaborate with other intercellular molecules to form lateral junctions. Thus, blocking integrin function could alter the junctional connection leading to permeation of macromolecules across endothelial monolayers.

Not only do focal adhesions maintain normal physiological endothelial barrier properties, their assembly and activity also mediate hyperpermeability responses under stimulated conditions in the presence of angiogenic factors, inflammatory mediators, or physical forces. For example, integrins are essential to the mechanotransduction of endothelial responses to shear stress, a well-known modulator of vascular permeability [79, 211, 514]. The $\beta_5$ subunit of integrins has been identified as a key molecule involved in the recruitment of kinases to focal adhesions in endothelial cells upon stimulation by vascular endothelial growth factor (VEGF) [29]. Mice deficient in integrin $\beta_5$ expression display a reduced vascular permeability response to VEGF treatment [123]. In addition, it has been shown that thrombin, a well-characterized hyperpermeability factor, contains an RGD-binding sequence that interacts with $\alpha_v\beta_3$ on endothelial cells leading to enhanced angiogenesis [197]. Most recently, an in vivo experiment demonstrated that plasma leakage across microvessels caused by fibrinogen degradation products is greatly attenuated in integrin $\beta_1$ knockout mice [170], further supporting the role of integrins in regulating endothelial hyperpermeability.

The precise mechanisms by which focal adhesions contribute to the maintenance of endothelial barrier function and hyperpermeability responses to stimulation are not completely understood [493]. It appears that various physical and chemical signals can be sensed and coordinated at the cell–matrix focal contact sites where integrins play a central role in transmembrane crosstalk between the cells and extracellular matrix. Within this context, focal adhesions are lipid raft domains containing scaffold proteins that bind to multiple intracellular signaling molecules (Figure 19). On one hand, ECM–integrin interactions induce "outside-in" signaling events that may contribute to the maintenance of endothelial barrier integrity. On the other hand, agonist-receptor binding triggers "inside-out" signaling events that modulate integrin–ECM adhesions. These responses alter the pattern or strength of integrin–ECM interactions, and in worse cases, cause disengagement of focal adhesions and even detachment of cells from ECM. Altered focal adhesions may also interfere with maintenance of normal cytoskeletal tension required for basal barrier properties. In response to certain stimuli, such as histamine, both the number and strength of integrin–matrix bonds are increased. This may contribute to hyperpermeability by transducing cytoskeletal contractile forces to cell–cell junctions leading to barrier opening.

The ECM also supports normal endothelial barrier function by preventing extravasation of circulating cells into the extravascular tissues [222, 490]. Under pathophysiological conditions, such as inflammation or metastatic cancer, activated leukocytes or invasive tumor cells secrete proteases and other enzymes capable of digesting ECM proteins and disengaging focal adhesions

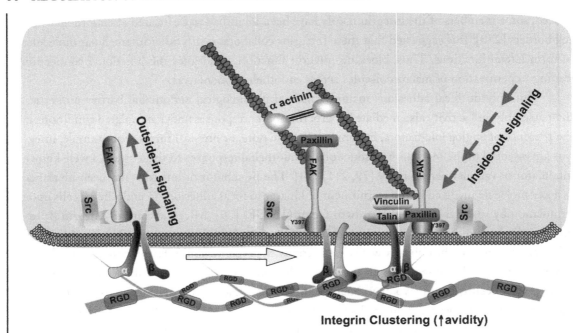

**Integrin Clustering (↑avidity)**

**FIGURE 19:** Outside-in and inside-out signaling at endothelial focal adhesions. Binding of RGD proteins to integrins triggers intracellular signaling events (outside-in signaling) including activation and recruitment of kinases (e.g., Src kinase, focal adhesion kinase (FAK). Kinase recruitment and activation increases integrin extracellular binding affinity and induces integrin clustering at focal adhesions (inside-out signaling). Clustering increases the collective binding capacity (avidity) of integrins at focal adhesions, strengthening adhesion to the basement membrane. Integrin clustering is further stabilized by intracellular connections to the actin cytoskeleton (via α-actinin, vinculin, talin, paxillin, etc.) and serves to maintain barrier integrity.

[43, 232, 384]. Certain inflammatory cells also bind to endothelial cell surface through integrins (e.g., leukocyte β2 integrins), which stabilizes leukocyte attachment to endothelium and facilitates transendothelial migration, enabling subsequent chemotactic migration across matrices. ECM breakdown permits migrating cells to enter the surrounding tissue [222]. A similar process occurs during angiogenesis, where ECM breakdown enables provisional cell migration along the leading edge of sprouting capillaries [189, 356].

## THE ENDOTHELIAL CYTOSKELETON
Similar to many other cell types, the endothelial cytoskeleton is composed of microtubules, intermediate filaments and actin filaments (Figure 20) [358, 414]. These structures are important for

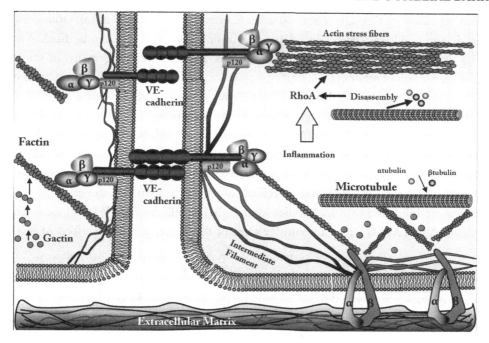

**FIGURE 20:** The endothelial cell cytoskeleton. Polymeric components of the cytoskeleton: actin microfilaments, microtubules and vimentin intermediate filaments, stabilize endothelial cell structure. Cell peripheral (cortical) actin filaments stabilize intercellular junctions and maintain focal adhesion integrity to support normal physiological endothelial barrier function. Microtubules and intermediate filaments provide additional support. Under inflammatory conditions, including RhoA activation, microtubules disassemble into $\alpha$- and $\beta$- tubulin subunits, and actin filaments, reorganize into parallel and linear stress fiber bundles that span the cell interior, pathological characteristics of endothelial hyperpermeability.

endothelial cell morphology, adhesion, and barrier function. While the structural support provided by all cytoskeletal components is important for barrier integrity, the actin cytoskeleton is most centrally important for regulation of endothelial permeability.

## Microtubules

Microtubules are tubular structures formed of polymers of heterodimeric subunits of alpha and beta tubulin (Figure 20) [299, 358]. The tubular structure is formed of 13 parallel polymeric filaments arranged in a ring. Microtubules are important for cell mitosis, morphology, and intracellular protein trafficking. In endothelial cells, microtubules are cross-linked to actin filaments and can affect endothelial permeability through effects on actin filaments. The stability of microtubules is determined by dynamic polymerization and depolymerization. Microtubules can be further stabilized by capping or by other posttranslational modifications. Dynamic rearrangements of microtubules affect the organization of other cytoskeletal components, and stabilization of microtubules is shown to protect endothelium against actin stress fiber formation and hyperpermeability [420]. Depolymerization of microtubules activates guanine nucleotide exchange factors, and signaling through Rho family GTPases, leading to actin stress fiber formation [48]. Signaling through Rho kinases has further consequences for endothelial barrier function (discussed in subsequent sections). Microtubules are destabilized by treatment of endothelial cells with the inflammatory cytokine TNF-α [418], or with thrombin [46], and may therefore contribute to barrier dysfunction in response to inflammation. Stabilization of microtubules in endothelial cells by cAMP [419], or activation of protein kinase A (PKA) [47], may account for some of the barrier protective effects of this kinase.

## Intermediate Filaments

Intermediate filaments are formed of heterogeneous polymeric protein arrangements [299, 474]. In most cells, the principle intermediate filament protein monomer is vimentin. Intermediate filaments are important for endothelial cell structure, and are expressed most abundantly in cells exposed to shear stress, e.g., aortic endothelial cells. Intermediate filaments are connected to cell–cell junction proteins and to focal attachments to the basement membrane (Figure 20). Intermediate filaments undergo structural rearrangements in response to shear stress and are known to transmit mechanical tension between cells joined by intercellular junctions [474]. The role of intermediate filaments in control of endothelial barrier function is not clear. It is believed that the principle function of intermediate filaments in endothelial cells is to provide redundant structural support and strength, stabilizing connections formed by the less resilient actin microfilaments [299]. For example, vimentin gene knock-out mice have no major vascular defects [89], indicating that vimentin filaments are not strictly necessary for normal physiological functions of the vascular endothelium. However, the connections of vimentin filaments to the cell junction protein VE–cadherin are disrupted during exposure to the hyperpermeability-inducing agent histamine [409], suggesting the potential involvement of intermediate filaments in stimulated permeability responses. Connections to focal adhesions suggest that vimentin filaments may play a role in organization or serve as scaffold for focal adhesion proteins. Alternatively, vimentin may be involved in integrin signaling at cell–matrix attachments.

## Actin Filaments

Actin filaments are linear polymers of filamentous (F)-actin [299, 358, 414]. The stability of actin filaments is dependent upon the concentration of globular (G)-actin within the cell (Figure 20).

Maintenance of the intracellular concentration of G-actin above the critical concentration (0.1 µM) favors actin polymerization and formation of actin filaments. Under normal physiological conditions, actin filaments are randomly distributed throughout the cell (short filaments and diffuse actin monomers) and at the cell peripheral band (cortical actin) [44, 299, 358]. Upon treatment of endothelial cells with hyperpermeability-inducing agents, such as thrombin or histamine, actin filaments organize into linear, parallel bundles across the cell interior (stress fibers). Stress fiber formation is often accompanied by a contractile cell morphology and formation of gaps between adjacent endothelial cells. In contrast, when endothelial cells are treated with barrier-protective agents such as sphingosine-1-phosphate, actin filaments re-organize and localize at the cell periphery, appearing to strengthen cell–cell contacts.

At least 80 proteins are known to bind actin and modify the organization or function of the actin cytoskeleton [299, 358]. Several proteins are known to regulate actin polymerization (cofilin, gelsolin, or 27-kDa heat shock protein) [413, 414]. In addition, the architecture of the actin cytoskeleton is modified by signaling molecules, i.e., Rho family small GTPases (Rac1, Cdc42, and RhoA), to form specialized structures, such as cortical actin of the cell periphery or actin stress fibers [44, 358]. Cytoskeletal rearrangement and stress fiber formation are known to induce endothelial hyperpermeability; however, the precise mechanisms and specific contribution of this process to permeability regulation are not well established. In most cases, cytoskeletal rearrangements are inseparable from other events such as actin–myosin binding and contraction.

## Actomyosin Contractile Machinery

Actomyosin contraction and increased cytoskeletal tension is a central mechanism for inducing endothelial hyperpermeability [145, 161, 320, 358, 413, 414]. Because endothelial cell junctions are connected to focal adhesions via the actin cytoskeleton, changes in cytoskeletal tension directly affect the barrier structure and function. In endothelial cells, as in muscle cells, myosin is bound to cytoskeletal actin, and cytoskeletal tension is increased by actin–myosin-based contractile activity (Figure 21). Actomyosin contractility is increased by phosphorylation of myosin regulatory light chain (MLC-2) [411, 413, 414]. Phosphorylation of MLC-2 causes an ATP-dependent change in the tertiary protein folding structure of myosin and a shift in position relative to actin. This shift produces actomyosin contractile force, increasing tension on the actin cytoskeleton. Because the actin cytoskeleton is connected both to cell junction proteins and to focal adhesions, the focal adhesion connection acts as a fulcrum, allowing cytoskeletal tension to physically pull apart cell–cell junctions and increase endothelial paracellular permeability. Conversely, dephosphorylation of MLC-2 decreases actomyosin contractility, relaxes the actin cytoskeleton, and decreases endothelial permeability. Thus, actomyosin contraction and actin cytoskeletal tension are proportional to the net phosphorylation status of MLC-2 (Figure 21).

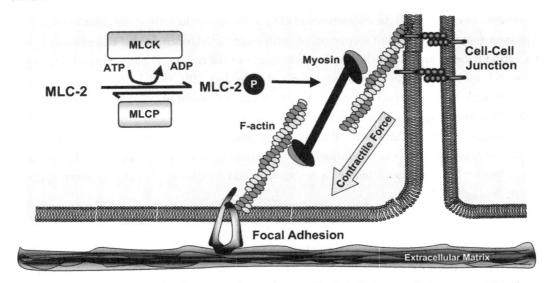

**FIGURE 21:** The endothelial cell contractile machinery. Endothelial paracellular permeability/hyper-permeability is controlled by actin-myosin driven cytoskeletal tension and retraction at cell–cell junctions. Actomyosin contraction is increased by phosphorylation of regulatory myosin light chain (MLC-2), which is phosphorylated by myosin light-chain kinase (MLCK) and dephosphorylated by myosin light-chain phosphatase (MLCP). The relative activities of MLCK and MLCP determine the net phosphory-lation status of MLC-2, and actomyosin contractility.

MLC-2 is phosphorylated by myosin light-chain kinase (MLCK) [145, 161] and is dephos-phorylated by myosin light-chain-associated phosphatase (MLCP) [464]. Normal physiological endothelial permeability is maintained and regulated by the activity of both MLCK and MLCP [413, 414]. The balance of MLCK/MLCP activity determines the steady-state phosphorylation status of MLC-2. Any stimulus that increases MLCK activity or decreases MLCP activity will in-crease MLC-2 phosphorylation and increase endothelial permeability. Many inflammatory agents and diseases cause endothelial hyperpermeability coinciding with or dependent on increased phos-phorylation of MLC-2 [145, 195, 509].

CHAPTER 5

# Signaling Mechanisms in the Regulation of Endothelial Permeability

The microvascular endothelial barrier is a dynamic system that undergoes constant remodeling with steady-state arrival on a moment-to-moment basis. Physical forces (such as shear stress) or biological factors (such as inflammatory mediators) cause changes in endothelial permeability by altering the synthesis and expression of cell–cell junction proteins or cell–matrix adhesion molecules over the long term (e.g., during chronic inflammation or angiogenesis), or by inducing conformational reorganization of the junction or focal adhesion complexes when stimulated by acute inflammatory agents, such as histamine, or during leukocyte transendothelial migration. Activated leukocytes also release reactive oxygen species and enzymes that further perpetuate and prolong barrier dysfunction. In microvessels, permeability responses are initiated by physical stimulation or agonist binding of endothelial surface receptors, followed by activation of a variety of intracellular signaling molecules, including second messengers, kinases, phosphatases, and GTPases. Signaling molecules control gene transcription, protein translation, posttranslational modification (phosphorylation, nitrosylation, glycosylation), vesicle trafficking (transcytosis), protein complex formation, lipid raft assembly/disassembly, and protein degradation (ubiquitination, proteolysis). These processes in turn control cell morphology, cytoskeleton integrity, contractility, and assembly/disassembly of cell junctions and focal adhesions, all of which can affect endothelial barrier function.

In the past two decades, our knowledge of the signaling events that regulate endothelial permeability has greatly increased. It is now realized that various kinases, phosphatases, GTPases, and the second messenger system operate together in functional units or networks. For example, focal adhesions are specialized microdomains that selectively recruit kinases and phosphatases. Recruitment of signaling molecules specifically to the site of activity allows them to function much more effectively than what is indicated by their overall concentration within the cell. In other words, recruitment can greatly enhance kinase specificity and activity locally at the recruitment site. Within focal adhesions, the focal adhesion kinase (FAK) has a powerful influence on integrin adhesion, cytoskeletal architecture, and signaling by growth factor receptors, much more so than at other locations within the cell. Because the complex interactions of intracellular signaling molecules vary according to cell types and environmental conditions, a new paradigm is emerging in the biology of cell signal transduction: signaling networks. Networks of signaling molecules interact with each other in patterns dictated by environmental conditions and, thus, can be mapped for specific diseases or chemical stimuli. Conceptually, this is easy to understand; however, in practice, it is exceptionally challenging to examine signaling networks in definitive ways or with causality. Although advances in bioinformatics and systems biology have led to the development of many sophisticated algorithms and computational models, most of our success in understanding cell signaling in disease has arisen from the reductionist approach of examining the behavior of subsets of signaling molecules, under a restricted set of experimental conditions, and endeavoring to understand the essential mechanistic events that mediate activity.

This chapter presents a summary of the signaling molecules that are well accepted as permeability regulators in microvascular endothelium. In general, the canonical pathway leading to increased permeability includes elevation of intracellular calcium and activation of protein kinase C (PKC), myosin light-chain kinase (MLCK), Src family kinases, and the small GTPase RhoA. Molecules that are considered barrier protectors (permeability-decreasing factors) include protein kinase A (PKA), Epac, sphingosine-1-phosphate (S1P), and the small GTPases Rac-1 and cdc42. Other molecules are known to alter barrier function but their effects (i.e., increase or decrease permeability) are controversial, including nitric oxide (NO), cyclic GMP, and Rac-1.

# INITIATION OF SIGNALING: ENDOTHELIAL RECEPTORS

Endothelial receptors are transmembrane proteins that specifically or selectively bind to extracellular ligands. Ligand engagement often induces a conformational rearrangement (e.g., dimerization) of the receptor and thereby relays signals to intracellular molecules that are readily bound, or newly recruited to the cytoplasmic domains of the receptor. Several types of receptors have been identified as frequent participants in signaling endothelial hyperpermeability, including receptor tyrosine kinases (RTKs), G-protein-coupled receptors (GPCRs), and integrin receptors.

## Receptor Tyrosine Kinases

Receptor tyrosine kinases (RTKs) are single-pass integral membrane proteins that bind to extracellular ligands and form dimers, inducing tyrosine kinase activity on the C-terminal cytoplasmic domain. RTKs contain SH2/SH3 homology recognition sites, and they recruit regulatory proteins and initiate intracellular signaling through protein tyrosine phosphorylation. The receptors of many growth factors, e.g., fibroblast growth factor (FGF) and VEGF, are RTKs.

VEGF is an angiogenic cytokine that was initially identified as a potent endothelial permeability-increasing factor [224, 261, 406]. VEGF-induced hyperpermeability is involved in tumor development, diabetic retinopathy, and ischemia/reperfusion injury [481]. VEGF binds to tyrosine kinase receptor subtypes VEGFR-1, -2 and -3 [32]. VEGFR-2, also known as KDR, is principally responsible for the hyperpermeability response to VEGF (Figure 22, center) [33]. VEGFR-2 is directly linked to phospholipase C-$\gamma$ (PLC$\gamma$). Upon binding of VEGF to VEGFR-2, PLC$\gamma$ becomes activated and catalyzes the conversion of $PIP_2$ to $IP_3$ and DAG. $IP_3$ binds to $IP_3$ receptors at the endoplasmic reticulum (ER) membrane inducing $Ca^{2+}$ release from the ER stores [41]. This internal $Ca^{2+}$ release stimulates subsequent opening of store-operated $Ca^{2+}$ (SOC) channels at the plasma membrane, inducing $Ca^{2+}$ influx. Elevated intracellular $Ca^{2+}$ triggers an array of downstream signaling events (discussed in a subsequent section). For example, working together with DAG, elevated $Ca^{2+}$ activates protein kinase C (PKC). VEGFR-2 activation results in activation of intracellular kinases including Src family kinases, MAP kinases, PI3 kinase [32], and protein kinase G [491]. These signaling events increase endothelial permeability. There is some evidence that VEGFR-1 also plays a role in the endothelial hyperpermeability response to VEGF, however, the contribution of VEGFR-1 appears to be minor compared to VEGFR-2.

## G-Protein-Coupled Receptors

Many permeability-increasing agents (e.g., histamine, thrombin, or bradykinin) signal through GPCRs [51, 86]. GPCRs are integral membrane proteins with seven transmembrane spanning domains and intracellularly coupled to heterotrimeric G-proteins (composed of $\alpha$, $\beta$, and $\gamma$ subunits) [351]. Upon binding to extracellular ligands, GPCRs induce activation of their respective G-proteins (e.g., subtypes $G_{12/13}$, $G_q$, $G_i$), characterized by exchange of GDP for GTP binding to the G-protein alpha subunit. The activated G-protein can then initiate downstream second messenger system. Each type of GPCR is coupled to a specific subtype of G-protein, which will initiate specific series of intracellular signaling events.

Histamine receptor subtypes $H_{1-4}$ are GPCRs that mediate the biological effects of histamine [343]. The $H_1$ receptor activates the G-protein subtype $G_q$ and induces endothelial hyperpermeability (Figure 22, right). The downstream of $G_q$ is phospholipase C (PLC$\beta$), which catalyzes the conversion of phosphatidylinositide-bisphosphate ($PIP_2$) into soluble inositol triphosphate ($IP_3$) and membrane-bound diacylglycerol (DAG). DAG and $IP_3$-induced increases in intracellular $Ca^{2+}$ induce hyperpermeability through similar mechanisms as those described for activation of VEGFR-2: $Ca^{2+}$ release from ER stores, opening of $Ca^{2+}$ channels at the cell membrane, $Ca^{2+}$ influx and subsequent activation of downstream kinases, including protein kinase G, MAP kinases, and MLCK (discussed in a subsequent section) [509].

Thrombin is a serine protease that cleaves fibrinogen into fibrin forming the base for blood clots during intrinsic or extrinsic (caused by injury) activation of coagulation. Thrombin also induces endothelial hyperpermeability by binding to protease-activated receptors (PARs) on the endothelial cell surface. PAR subtypes 1, 3, and 4 are GPCRs that mediate the effects of thrombin binding [51, 247]. Thrombin binding to PAR-1 on endothelial cells activates G-protein subtypes: $G_{12/13}$, $G_q$, and $G_i$ (Figure 22, left) [300, 460]. At low concentrations of extracellular thrombin, binding to PAR-1 activates $G_{12/13}$ leading to RhoA kinase activation, increasing actomyosin contractility and actin stress fiber formation, thereby increasing endothelial permeability. At higher concentrations of thrombin, similar to the effects of histamine on $H_1$, PAR-1 activation of $G_q$ increases intracellular $Ca^{2+}$ concentration through activation of $Ca^{2+}$ channels. PAR-1 also induces $G_i$ activation causing inhibition of adenylate cyclase, which contributes to hyperpermeability by preventing formation of cyclic adenosine monophosphate (cAMP) and suppressing barrier-protective effects of protein kinase A (PKA).

Bradykinin is produced during inflammation or ischemia/reperfusion injury following cleavage of the precursor kinogen by kallikrein. Occupancy of the bradykinin $B_2$ receptor on microvascular endothelial cells increases cytosolic $Ca^{2+}$ via similar mechanisms to those of the histamine

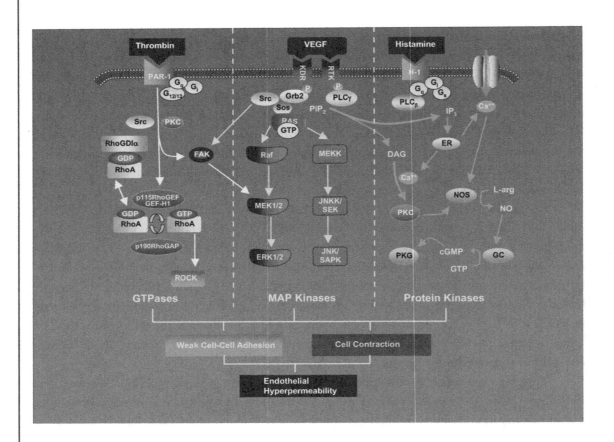

$H_1$ receptor or PAR-1, and activates membrane-bound phospholipase A2 ($PLA_2$) to produce arachidonic acid (AA) [364]. AA metabolism by lipoxygenase and cyclooxygenase (COX) produces leukotrienes, prostanglandins, and thromboxin A2. Most of these metabolites have been shown to cause endothelial dysfunction and promote leukocyte diapedesis. The signaling mechanisms involved in their actions include increased cytosolic $Ca^{2+}$, reactive oxygen species (ROS) production, and activation of nitric oxide synthase (NOS) and protein kinase C (PKC).

## Integrin Receptors

Integrins are transmembrane glycoproteins, dimers of alpha and beta subunits that bind extracellular matrix ligands or to immunoglobin (Ig) superfamily adhesion molecules [125]. Integrins have short intracellular C-terminal domains that are anchored to the actin cytoskeleton, serving as a signaling platform to recruit or interact with multiple intracellular molecules (Figure 19) [177, 301]. Integrins conduct "outside-in" signaling, where integrin binding to extracellular ligands triggers

---

**FIGURE 22:** Receptor-mediated intracellular signaling contributing to endothelial hyperpermeability (adapted from P. Kumar et al. 2009 *Expert Reviews in Molecular Medicine*, 11(e19): 1–20). VEGF signaling is mediated through binding and activation of receptor tyrosine kinases (RTKs), including KDR (VEGFR2) (center). Tyrosine kinase activation recruits intracellular adapter proteins (Grb2, SOS) and activates multiple protein kinase-dependent (Src kinase,) intracellular signaling cascades (initiated by phospholipase C-gamma (PLCγ), Ras-Raf-MAPK, or Src kinase). Other hyperpermeability-inducing agents such as thrombin or histamine act through activation of G-protein ($G_i$, $G_q$, $G_{12/13}$, or $G_s$)-coupled receptors (left and right, respectively). G-proteins selectively activate several intracellular signaling cascades mediated through Src kinase, protein kinase C (PKC) or inositol triphosphate ($IP_3$)-mediated release of intracellular $Ca^{2+}$ and subsequent activation of plasma membrane $Ca^{2+}$ channels. Signaling through these receptors converges several intracellular signaling pathways known to mediate hyperpermeability through effects on cell adhesion, cytoskeletal morphology and actin-myosin mediated cell contraction. Major pathways involved in receptor-mediated hyperpermeability involve small GTPases (Ras, Rho), MAPK/ERK kinases and other protein kinases (PKC, PKG). Additional abbreviations: Src family kinase, or *src* gene product (Src); focal adhesion kinase (FAK); guanosine diphosphate (GDP); guanosine triphosphate (GTP); Rho GDP-dissociation inhibitor 1 (Rho-GDI); 115 kDa guanine-nucleotide-exchange factor (p115RhoGEF); p190 Rho GTPase-activating protein (p190RhoGAP); Rho kinase (ROCK); growth factor receptor-bound protein 2 (GRB2); Son of sevenless (Sos); *ras* gene product (Ras); extracellular regulated kinase (ERK); mitogen-activated protein kinase/ERK kinase (MEK); stress-activated protein kinase (SAPK); SAPK/ERK kinase (SEK); c-Jun N-terminal kinase (JNK); diacylglycerol (DAG); endoplasmic reticulum (ER); L-arginine (L-Arg); nitric oxide (NO); nitric oxide synthase (NOS); guanylate cyclase (GC); cyclic guanosine monophosphate (cGMP).

intracellular signaling events, including activation of kinases (e.g., FAK, Src), Rho GTPases, and cytoskeletal linker proteins (e.g., paxillin, talin, vinculin, $\alpha$-actinin), as well as cell signaling cross-talk with growth factor receptors. Outside-in signaling occurs through a ligand-induced change in protein conformation of alpha and beta subunits, which causes the integrin C-terminal domains to be separated from each other [233]. Separation of the C-terminal domains allows for interactions with the aforementioned intracellular signaling molecules. Integrin outside-in signaling reinforces endothelial focal adhesion connections to the actin cytoskeleton, and maintains endothelial barrier function. Disruption of integrin binding to the ECM, or integrin binding to ligands (growth factors, etc.) released by the ECM, triggers intracellular signaling events leading to endothelial hyperpermeability.

Integrins also participate in "inside-out" signaling, where agonist-initiated intracellular signaling events alter the integrin–ECM binding probability, affinity, or avidity [177, 235, 286, 448]. For example, in many cell types (e.g., platelets, leukocytes), intracellular binding of integrin to talin induces a structural rearrangement of $\alpha$ and $\beta$ integrin subunits with respect to each other and increases the extracellular binding affinity [4, 235]. Increased integrin affinity is believed to result from a shift in protein structural conformation of the alpha subunit, enabling salt bridge formation with the beta subunit, and stronger ligand binding [286, 417]. In neutrophils, normal integrin association with talin is disrupted by integrin activation [4, 235]. This enables integrin lateral diffusion and clustering within the cell membrane. Integrin clustering allows integrins to more readily bind to multivalent extracellular ligands due to the close proximity of binding sites on multiple integrin molecules (increased valence; avidity). In endothelium, integrin clustering at focal adhesions increases binding to the ECM. In neutrophils, integrin inside-out signaling contributes to endothelial barrier dysfunction by increasing neutrophil binding to microvascular endothelium and facilitating transendothelial migration.

The leukocyte-specific ($\beta_2$) integrins facilitate leukocyte adhesion to endothelium under inflammatory conditions [125, 126, 192, 233, 235, 268]. During inflammation, chemokines (growth factors, interleukins, etc.) released by endothelial cells or displayed on the endothelial cell surface can bind to leukocyte cell–surface chemokine receptors. For example, interleukin-8 (IL-8) attracts leukocytes when displayed on the luminal surface of endothelial cells. In response to inflammatory conditions, IL-8 is moved in vesicles from the basolateral to the apical surface of endothelium for presentation to leukocytes in the blood circulation [307]. To complement endothelial chemokine activity, activated leukocytes may also increase surface expression of chemokine receptors [125]: G-protein-coupled receptors [521] that trigger multiple intracellular signaling events in endothelial cells. Chemokine binding to chemokine receptors initiates inside-out signaling to integrins, which increases integrin avidity (discussed above) and adhesion to endothelial cell surface proteins.

The leukocyte $\beta_2$ integrins[1]: lymphocyte function antigen 1 (LFA-1; $\alpha_L\beta_2$), Mac-1 ($\alpha_M\beta_2$), and p150,95 ($\alpha_X\beta_2$) bind to endothelial ICAM-1; $\alpha_D\beta_2$ binds to vascular cell adhesion molecule 1 (VCAM-1) [125, 126, 166, 268]. $\beta_2$ integrins are central to leukocyte firm adhesion; however, in some tissue types, additional leukocyte surface proteins bind to endothelial surface proteins, including $\alpha_4$ integrins ($\alpha_4\beta_1$ and $\alpha_4\beta_7$). Firm adhesion mediated by leukocyte integrins, and other adhesion molecules, is necessary for leukocyte transendothelial migration (diapedesis). Adhesion molecule binding to leukocyte integrins triggers additional outside-in signaling, leading to leukocyte activation and secretion of chemical factors (reactive oxygen species, NO, eicosanoids, leukotrienes, prostaglandins, proteases, etc.) that cause endothelial hyperpermeability [55].

# INTRACELLULAR SIGNAL TRANSDUCTION: SECOND MESSENGERS
## Calcium

Calcium is a ubiquitous signaling molecule with multiple roles in signaling endothelial hyperpermeability [1, 444]. $Ca^{2+}$ is a cofactor for $Ca^{2+}$-dependent enzyme activity throughout the cell, including calmodulin, PKC, NOS, and many others. Affecting numerous cellular functions, the intracellular-free $Ca^{2+}$ concentration is very tightly and rapidly regulated. Excess $Ca^{2+}$ is sequestered in intracellular compartments, including mitochondria and the endoplasmic reticulum (ER). Free intracellular $Ca^{2+}$ arises upon release from ER or influx through cell membrane $Ca^{2+}$ channels (Figure 23). In the cytoplasm, $Ca^{2+}$ binds to numerous $Ca^{2+}$ binding proteins, including calmodulin, which increases the activity of several enzymes, including MLCK. In this example, $Ca^{2+}$-dependent activation of MLCK increases cytoskeleton contractility and endothelial permeability.

The majority of sequestered $Ca^{2+}$ in endothelial cells is retained in the ER. ATP-driven sarcoplasmic/endoplasmic reticulum $Ca^{2+}$ (SERCA) pumps actively transport residual $Ca^{2+}$ from the cytoplasm into the ER lumen (Figure 23) [1, 444]. Thus, under normal physiological conditions, the ER contains a great reserve of intracellular "stores" $Ca^{2+}$ that can be rapidly released as needed by the cell. The principle mechanism for ER release of $Ca^{2+}$ is through the $IP_3$ receptor-gated $Ca^{2+}$ channels. In general, $IP_3$ is produced as a hydrolysis product of activated PLC at the cell membrane. Once accumulated in the cytosol, $IP_3$ binds to the $IP_3$ receptor on the ER membrane and triggers release of $Ca^{2+}$ from the ER store into the cytoplasm.

---

[1]The International Union of Immunological Societies/World Health Organization nomenclature designates leukocyte integrins as Clusters of Differentiation (CD): CD18 ($\beta_2$), CD11a ($\alpha_L$), CD11b ($\alpha_M$), CD11c ($\alpha_X$) and CD11d ($\alpha_D$). CD nomenclature is characterized and maintained by Human Cell Differentiation Molecules (www.hcdm .org).

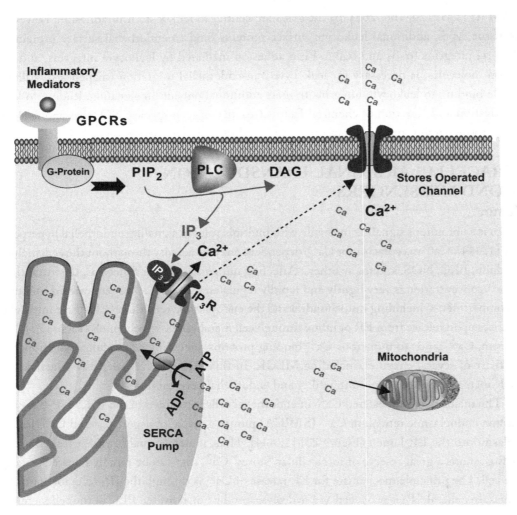

**FIGURE 23:** Calcium signaling through G-protein-coupled receptors (GPCRs). G-protein mediated activation of phospholipase C (PLC) catalyzes the conversion of phosphatidylinositol bisphosphate ($PIP_2$) into diacylglycerol (DAG) and inositol triphosphate ($IP_3$). $IP_3$ binding to the endoplasmic reticulum (ER) $IP_3$ receptor ($IP_3R$) causes intracellular $Ca^{2+}$ release from the ER. Activation of the $IP_3R$ triggers activation of plasma membrane $Ca^{2+}$ channels, and $Ca^{2+}$ influx dramatically raises the intracellular-free $Ca^{2+}$ concentration ($Ca^{2+}$-induced $Ca^{2+}$ influx) to induce $Ca^{2+}$-dependent cell signaling events and endothelial hyperpermeability. Excess intracellular free $Ca^{2+}$ is transported into the ER by the ATP-driven sarco(endo)plasmic reticulum $Ca^{2+}$ (SERCa) pump, or is internalized by mitochondria.

Depletion of ER $Ca^{2+}$ in turn activates $Ca^{2+}$ channels located at the cell membrane [444]. This mechanism of $Ca^{2+}$ channel activation is referred to as $Ca^{2+}$-induced $Ca^{2+}$ influx [41], also known as store-operated $Ca^{2+}$ influx. In brain endothelial cells, the principle store-operated $Ca^{2+}$ (SOC) channel is the transient receptor potential $Ca^{2+}$ (TRPC) channel [444]. TRPC channel-mediated calcium spikes may participate in the endothelial hyperpermeability response to certain agonists via disruption of cell junction complexes or cytoskeletal reorganization. TRPC channel activation is also accompanied by MLC phosphorylation. However, it is not clear whether MLC phosphorylation is a cause or consequence of TRPC activation because investigators have shown that TRPC inhibitors also prevent MLC phosphorylation, and conversely, that MLCK inhibitors prevent TRPC activation.

It is also shown in the case of VEGF-induced hyperpermeability that DAG production in response to PLC activation can directly activate TRPC channels (i.e., store-independent isoforms TRPC3/6), and increase intracellular $Ca^{2+}$ [355]. In this case, VEGFR-2, PLCγ, TRPC6, and other signaling molecules, including DAG, endothelial nitric oxide synthase (eNOS), and protein kinase C isoform alpha (PKC-α) are localized in caveolin-1-rich lipid raft microdomains within the plasma membrane [32]. Signaling within raft microdomains is extremely efficient because of the proximity of receptors, effectors, and signaling intermediates.

## cAMP

The second messenger cAMP improves endothelial barrier function and protects microvasculature from hyperpermeability [458, 496, 509]. Endothelial intracellular cAMP production is increased in response to $β_2$-adrenergic agonists (e.g., isoproterenol or forskolin) or the cell-permeable cAMP analog, 8Br-cAMP. The protective effects of cAMP are prevented by the endogenous PKA inhibitor PKI, suggesting that barrier protection is mediated through activation of PKA [344, 509]. PKA stabilizes cell–matrix adhesions and actin filaments, dissociates myosin from F-actin, and decreases actomyosin contractility. Decreased actomyosin contractility is caused by PKA inhibition of MLCK activity [413, 510], while cytoskeletal stabilization is likely due to PKA-mediated inhibition of RhoA [460].

The regulated production of cAMP may involve differential activation of adenylate cyclase (AC) isoforms [82, 482]. Elevated intracellular $Ca^{2+}$ inhibits AC5 and AC6 and dramatically decreases production of cAMP in microvascular endothelial cells. The intracellular concentration of cAMP is also regulated by the activity of phosphodiesterase isoforms (PDE1–5), enzymes that hydrolyze cyclic nucleotides [226]. The activity levels of various PDE isoforms can be increased or decreased by intracellular signaling molecules, including cGMP. In endothelial cells, cGMP has been shown to activate PDE2 or PDE4, or inhibit PDE3, with reciprocal effects on cAMP concentration [496].

## NO-cGMP

Nitric oxide is a short-lived free radical produced in response to phosphorylation and activation of nitric oxide synthase (NOS) (Figure 24). Three isoforms of NOS have been implicated in inflammatory responses: endothelial (eNOS; NOS1), inducible (iNOS; NOS2), and neuronal (nNOS; NOS3) [407]. Activated NOS catalyzes the conversion of the amino acid L-arginine (Arg), to citrulline and NO [239, 314]. This reaction is inhibited by chemical analogs of L-arginine (e.g., $N^G$-monomethyl-L-arginine (L-NMMA) or $N^{\omega}$-nitro-L-arginine-methyl ester (L-NAME)) [118]. NO is highly reactive with numerous proteins through posttranslational modifications: NO binds to cysteine (Cys) residues on proteins to form $S$-nitrosothiol groups (i.e., $S$-nitrosylation); NO combines with the oxygen-free radical, superoxide ($O_2{}^{\cdot}$) to form the highly reactive species peroxynitrite (ONOO$^-$) [55, 239, 314, 331, 367]. Of great significance to cell viability, ONOO$^-$ interferes with the activity of mitochondrial cytochrome C and disrupts oxidative phosphorylation [314]. This impairs energetic metabolism and prevents $Ca^{2+}$ sequestration in mitochondria, normally accounting for 25% of sequestered $Ca^{2+}$ within the cell (Figure 24) [103].

The role of NO in regulating endothelial barrier function is controversial [118, 244, 246, 274, 508]. NO has opposing effects on endothelial permeability depending upon the endothelial model system, species, or vascular bed examined. NO is a potent inducer of hyperpermeability in coronary microvessels [491, 514] and other microvascular beds [244]. However, NO has barrier-protective effects in cultured endothelial cells and in some organs, such as the skin, kidney, or intestine [246, 253]. Because NO is a vasodilator in resistance vessels, NO may affect transendothelial flux of fluid in vivo by modulating blood flow in some vascular beds [244, 274]. Furthermore, NO is known to inhibit platelet aggregation and leukocyte adhesion, processes that indirectly affect plasma transport across microvessels. Thus, hemodynamics must be considered when interpreting the effects of NO in vivo. Also, the effects of NO on permeability in cultured cells derived from macrovascular (e.g., aorta) or nonexchange microvascular (e.g., arterioles) origin are not representative of the effects of NO in true exchange microvasculature and must be interpreted cautiously. That said, the basis for many of the observed differences in the effects of NO are attributed to the activity of different NOS isoforms. For example, the hyperpermeability-inducing effects of VEGF treatment are attenuated in eNOS (–/–) mice, but not in iNOS (–/–) mice [141]. In addition, subcellular localization of eNOS may influence the effects of eNOS on endothelial permeability [118]. In response to certain stimuli, eNOS is released from the plasma membrane and moves to other locations within the cell; release of eNOS is attributed to loss of a palmitoyl moiety (a lipophilic membrane anchor). Treatment with the hyperpermeability-inducing agents VEGF or platelet activating factor (PAF) causes eNOS to relocate from the plasma membrane to the cytosol [392–394]. In contrast, treatment with the vasodilator acetylcholine causes eNOS to be sequestered to the Golgi complex. Because NO is a short-lived and unstable molecule in physiological tissues, proximity to target molecules has a strong influence on the regulatory effects of NO.

Most of the existing evidence for the permeability-inducing effects of NO in microvasculature comes from experimental studies employing pharmacological NOS isoform-selective inhibitors and chemical NO donors that act broadly throughout the entire cell. Treatment of microvascular endothelium with chemical NO donors (e.g., *S*-nitroprusside) induces hyperpermeability, a response dependent on guanylate cyclase (GC) activity (Figure 24) [515]. The mechanism for NO activation

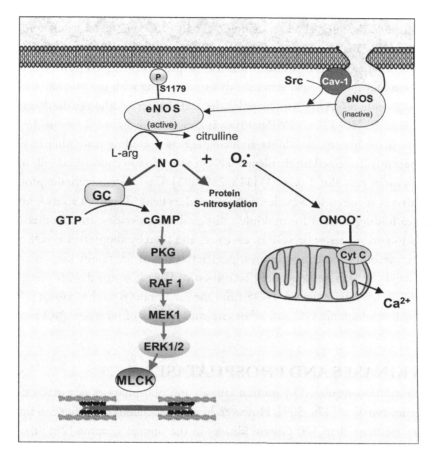

**FIGURE 24:** Endothelial hyperpermeability signaling through nitric oxide. Activation of endothelial nitric oxide synthase (eNOS) by $Ca^{2+}$ binding or through phosphorylation of serine (S) 1179, catalyzes the conversion of L-arginine to citrulline and nitric oxide (NO). Under inflammatory conditions, eNOS activation may occur in response to Src-mediated phosphorylation of caveolin-1, and subsequent release of sequestered eNOS from caveolae. NO production causes microvascular endothelial hyperpermeability by combining with superoxide-free radicals to form peroxynitrate ($ONOO^-$) and impairing mitochondrial function. NO also activates guanylate cyclase (GC), which converts GTP to cGMP and activates protein kinase G (PKG). PKG activates the Raf-MEK-ERK signal cascade and causes endothelial hyperpermeability through increased actomyosin contraction.

of GC is not known, however, *S*-nitrosylation of GC is a possibility [118]. Once activated, GC catalyzes the conversion of guanosine triphosphate (GTP) into cyclic guanosine monophosphate (cGMP). Increased cGMP levels in the cell in turn activate cGMP-dependent protein kinase G (PKG). Activation of PKG by treatment with the cGMP analog compound 8-bromo-cGMP induces hyperpermeability in isolated coronary venules, and hyperpermeability is prevented by treatment with the GC inhibitor LY83503 [515].

In endothelial cells, intracellular signaling through PKG is not completely understood; however, hyperpermeability may be mediated through activation of ERK1/2 and MLCK (Figure 24); inhibitors of these kinases also prevent microvessel hyperpermeability [497]. Alternatively, increased cGMP production may induce hyperpermeability by competing with the protective effects of cyclic adenosine monophosphate (cAMP) (discussed in the next chapter). Although the downstream mechanisms are not clear, the NO-GC-cGMP pathway is known to mediate microvessel hyperpermeability in response to many hyperpermeability-inducing agents, including histamine and VEGF [509].

A more recently described mechanism of eNOS activation in endothelial cells involves release of eNOS inhibition by caveolin-1 (Cav-1) [118, 239, 393]. Cav-1 is a membrane protein associated with caveolae that is required for vesicle-mediated endocytosis (discussed above); however, Cav-1 also acts as a scaffolding protein for multiple cell signaling molecules. In endothelial cells, Cav-1 normally binds to and sequesters eNOS in caveolae, and thereby suppresses eNOS activity [220]. Several hyperpermeability-inducing stimuli trigger activation of Src family kinases and subsequent activation of Cav-1. Src kinase-dependent activation of Cav-1, coupled with phosphorylation of eNOS at Ser-1179, releases activated eNOS from the membrane into the cytosol [290]. Activated eNOS in turn activates soluble GC and other enzymes by way of increased NO production (Figure 24) [118].

## PROTEIN KINASES AND PHOSPHATASES

Protein phosphorylation regulated by protein kinases and phosphatases is a major determinant of endothelial barrier function [496, 509]. However, kinase-dependent signaling can be exceptionally complex. There are more than 500 protein kinases in the human genome [291] that catalyze the transfer of inorganic phosphate supplied by ATP onto serine (Ser; S), threonine (Thr; T) or Tyrosine (Tyr; Y) amino acid residues of numerous proteins. The specificity of protein phosphorylation is determined by the local amino acid sequence (consensus sequence) surrounding the S/T/Y residue to be phosphorylated [10]. Consensus amino acid sequences are unique recognition sequences for phosphorylation by individual kinases (e.g., protein kinase A, protein kinase C). Phosphorylation consensus sequences are specific for recognition by a particular kinase, yet in many cases may be phosphorylated in a nonspecific fashion by other kinases, depending upon the local chemical environment and the relative abundance of active kinases.

Protein phosphorylation is opposed by protein phosphatase-catalyzed dephosphorylation. Many protein phosphatases are promiscuous, in that they do not strictly require amino acid consensus sequences to identify substrates for dephosphorylation. For example, protein phosphatase 1 (PP1) is a Ser/Thr directed phosphatase, which will remove phosphates from Ser/Thr residues on numerous phosphorylated proteins. Other phosphatases dephosphorylate a specific phosphoprotein. For example, myosin light-chain phosphatase (MLCP) dephosphorylates myosin light chain (MLC) [179].

In most regulatory schemes, phosphorylation of a protein is determined by the relative activity of a kinase and an opposing phosphatase. The phosphorylation state is increased with increased kinase or decreased phosphatase activity. Conversely, phosphorylation is decreased with decreased kinase or increased phosphatase activity. Thus, the activity of a kinase or phosphatase per se may not determine the phosphorylation status of a target protein. One must either examine the activity of both the kinase and the opposing phosphatase or measure the substrate phosphorylation level to make this determination. Furthermore, the abundance of a kinase does not necessarily dictate activity. Many kinases exist in active and inactive states. Activation occurs through phosphorylation or release of autoinhibition. Once in the activated state, a kinase can bind to ATP and acquire the capacity to transfer phosphate (activity). Other factors may enhance or interfere with kinase activity, such as abundance of cofactors (e.g., $Ca^{2+}$, $Mg^{2+}$), binding to inhibitory proteins, or recruitment to proximity of phospho-accepting substrates by scaffolding proteins. In addition, sub-cellular localization must be considered in analyzing kinase activity and kinase-dependent cell signaling. With respect to MLCK signaling in regulating endothelial barrier function, its interaction with the actin cytoskeleton is the most important factor. On the other hand, when protein kinase C is activated, it translocates from the cytosol to the cell membrane; thus, its phosphorylation activity at cell–cell junctions becomes an important area of investigation [150].

## Myosin Light-Chain Kinase

Myosin light-chain kinase (MLCK) is a family of soluble protein kinases that phosphorylate the regulatory myosin light chain (MLC-2) [413]. MLCK generally recognizes the consensus sequence R/K$_3$-X$_2$-R-X$_2$-S (arginine (R); Lysine (K); X = any amino acid) though recognition may also be dependent upon three-dimensional protein conformation [10]. It is not known whether MLCK phosphorylates regulatory proteins other than MLC, as suggested by some investigators. Nonetheless, phosphorylation of MLC-2 by MLCK is a central molecular mechanism regulating endothelial permeability and barrier function [413].

There are multiple isoforms of MLCK [74, 256, 413, 478]. Skeletal muscle and cardiac muscle MLCK are expressed only in skeletal or cardiac muscle cells, respectively. Smooth muscle and nonmuscle MLCK isoforms (smMLCK and nmMLCK, respectively) have wide tissue distribution, and both are expressed in microvascular endothelial cells. Because of this, it is exceptionally difficult to distinguish between the activities attributed to smMLCK vs. nmMLCK isoforms. This is important because the regulation of these isoforms is quite different, and they are likely to play different regulatory roles within microvascular endothelium.

The smMLCK and nmMLCK isoforms are derived from the same gene (*mylk1*) and, therefore, have identical amino acid sequences and protein structures [256]. Both nmMLCK and smMLCK have identical structures responsible for ATP binding, kinase activity, and binding to MLC. The nmMLCK isoforms (MLCK 1, 2, 3a, 3b and 4) are much larger (approximately 210 kDa) than smMLCK (108 kDa) due to an N-terminal deletion (posttranscriptional splicing variation) of 922 amino acids that is absent from smMLCK. This stretch of amino acids contains multiple sites for phosphorylation and regulation by other protein kinases (e.g., PKC, PKA, MAP kinases). There are also differences in regulation between the nmMLCK isoforms [45, 146]. For example, MLCK1 and MLCK3a are phosphorylated by p60 Src family kinases at two tyrosine residues (Tyr-464 and Tyr-471) that are not present in other MLCK isoforms. More recent studies suggest that there are differences in the expression and subcellular localization of MLCK isoforms, which may serve as a basis for differential regulation of the activity of individual MLCK isoforms, or regulation of specific cellular processes by MLCK isoforms [115]. The most abundant MLCK isoforms in endothelial cells are MLCK1 and MLCK2 [256]; however, the difference in regulatory activity of these two isoforms is not well understood.

Calmodulin (CaM) is a major regulator of MLCK activity. MLCK contains a CaM binding domain that maintains MLCK in an inhibited (inactive) state when CaM is not bound. CaM binding to MLCK releases autoinhibition of kinase activity, allowing MLCK to become activated. In endothelial cells, CaM binding to MLCK in the presence of $Ca^{2+}$ is necessary for MLCK activity, but alone is insufficient to activate MLCK [146, 463]. Endothelial MLCK activation requires additional signaling events. Once activated, MLCK induces phosphorylation of MLC at serine-19 and, subsequently, at threonine-18, which causes a shift in the tertiary protein structure of MLC and increases actomyosin contractility (Figure 25).

MLCK activity can be inhibited by pharmacological compounds, such as ML-7 or ML-9. Treatment with ML-7 or ML-9 decreases albumin permeability in coronary microvessels. The drugs also reduce the basal microvascular permeability in the gut or mesentery [516]. This is consistent with the notion that tonic MLCK activity is present under physiological conditions that maintains normal resting endothelial barrier function. In support of this observation, treatment with the protein phosphatase inhibitor calyculin-A also increases MLC phosphorylation, indicating that there is residual MLCK activity in unstimulated microvessels. Therefore, under normal conditions in the absence of stimulation, basal endothelial permeability is maintained by tonic MLCK kinase activity.

It should be noted that the effects of ML-7 on unstimulated microvascular endothelium are rarely seen in cultured endothelial cell monolayers [413]. As discussed in the preceding chapter,

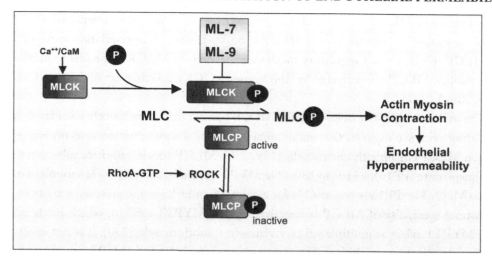

**FIGURE 25:** Myosin light-chain kinase (MLCK)-dependent endothelial hyperpermeability. Endothelial actomyosin contraction is dependent upon MLCK-mediated phosphorylation of regulatory myosin light chain (MLC). MLCK activation occurs through $Ca^{2+}$-dependent calmodulin (CaM) binding to MLCK, and/or through phosphorylation of MLCK by protein kinases (e.g., Src, PKC). Phosphorylation of MLC by MLCK is opposed through dephosphorylation by activated myosin light-chain phosphatase (MLCP). Inactivation of MLCP through phosphorylation by Rho kinase (ROCK) (downstream of RhoA-GTP) prevents dephosphorylation of MLC, resulting in increased MLCK-dependent net phosphorylation of MLC. The phosphorylation status of MLC directly determines actomyosin contractility and endothelial permeability/hyperpermeability. MLCK-dependent signaling can be tested with ML-7 or ML-9, compounds that selectively inhibit MLCK activity.

cultured endothelial cells lack critical elements that maintain physiological permeability in intact microvessels, i.e., glycocalyx, basement membrane, and surrounding matrix and cells. In addition, hyperpermeability in intact microvessels is mediated by selective opening of cell–cell junctions near less than 5% of endothelial cells in the microvessel wall [96, 97]. Even within the same microvessel endothelium, heterogeneity exists between adjacent endothelial cells [97]. Therefore, cultured endothelial cells may not contain the appropriate phenotype for maintenance of normal physiological barrier properties; these cells more likely represent injured or inflammatory phenotypes.

## Myosin Light-Chain Phosphatase

Myosin light-chain phosphatase (MLCP) activity opposes MLCK activity through dephosphorylation of MLC-2 (Figure 25) [413, 414]. The phosphorylation status of MLC-2 results from the dynamic balance of MLCK activity and opposing MLCP activity. In the presence of MLCK activity, phosphorylation and inactivation of MLCP increases the net phosphorylation status of MLC-2. Thus, MLCP inhibition is a regulatory mechanism leading to MLCK-dependent endothelial hyperpermeability. MLCP is inactivated by Rho kinase (ROCK), downstream of the Rho family small GTPase, RhoA [247].

The molecular identity of endothelial MLCP has not been conclusively elucidated; however, accumulating evidence suggests that endothelial MLCP is the same molecule as myosin phosphatase (a.k.a. MLCP) in smooth muscle cells [179, 276]. MLCP consists of three subunits: a 38-kDa type 1 phosphatase (PP1δ), a 110- to 130-kDa MYPT1 subunit, and a 20-kDa subunit of unknown function (M20). The PP1δ is responsible for catalytic serine/threonine protease activity of MLCP. The substrate specificity of MLCP is conferred by the MYPT1 subunit, which binds directly to myosin. MYPT1 exists as multiple splice variants in smooth muscle [230]; it is not known which MYPT1 variant is responsible for MLCP activity in endothelium. MLCP is inactivated through phosphorylation of MYPT by Rho kinase (ROCK) at Thr-696 and Thr-853 [130, 321], which dissociates MLCP from myosin [462]. In endothelial cells, MLCP activity is inhibited by CPI-17, an endogenous specific inhibitor of MYPT [111, 238]. CPI-17 augments MLCK-dependent endothelial hyperpermeability in response to histamine and other agonists.

## Src Family Kinases

Src family kinases are nonreceptor tyrosine kinases that perform multiple signaling functions in many cell types. The roles of Src family kinases are determined by differential expression of Src kinase isoforms in specific cell types and by intracellular localization with other signaling molecules within the cell.

The prototypical src isoform, viral src kinase (v-src), was originally isolated from the cancer-inducing Rous sarcoma virus [61, 365]. The Src family kinases comprise at least 11 isoforms: Blk, Brk, Fgr, Frk, Fyn, Hck, Lck, Lyn, Src, Srm, and Yes [303, 440] with molecular weights ranging 52–62 kDa [202]. The isoforms Src, Fyn, and Yes are widely expressed in mammalian tissues [335] and are known to be involved in endothelial cell signaling. Other isoforms have relatively more restricted tissue distribution, mainly in blood cells or immune cells, and they may be important in endothelial barrier dysfunction in that they are expressed in inflammatory cells (neutrophils, monocytes and macrophages) and tumor cells that can transmigrate across endothelium.

The regulatory activity of Src family kinases is based on highly conserved protein structure, with eight functional domains [202, 282]. The N-terminal domain contains a glycine site for myristoylation, a lipophilic modification that anchors Src kinases to the membrane and can determine partitioning within the membrane into lipid raft microdomains. Src kinases also contain four Src homology (SH) domains (SH1–4). The SH4 domain is adjacent to the N-terminal domain. The

SH4 domain in some Src family kinases contains a palmitoylation site, which also serves as a membrane anchor, similar to the myristoylation site. Juxtaposed to the SH4 domain is a unique region that is different for each Src family isoform. Following this unique region are an SH3 domain and an SH2 domain. SH3 and SH2 domains contain binding sites for SH3/SH2 protein–protein interactions. SH2/SH3 domains allow Src family kinases to bind to scaffolding proteins, receptors, and protein aggregates within cells that contain complementary SH2/SH3 domains. Selective binding and localization within the cell imparts specificity to cell signaling events that depend upon Src family kinases. The SH2 domain is connected, via a linker sequence, to the kinase catalytic domain. The catalytic domain is responsible for tyrosine kinase activity of Src family kinases, where autophosphorylation of Tyr-416 (v-Src numbering) within the catalytic domain is required for tyrosine kinase activity. At the C-terminal end of Src kinases is a regulatory domain. Phosphorylation of Tyr-527 within the regulatory domain by carboxy-terminal Src kinase (Csk) suppresses tyrosine kinase activity [59, 440]. Dephosphorylation of Tyr-527 causes a change in the protein folding conformation of Src, which permits autophosphorylation of Tyr-416, and is therefore required for Src kinase activation.

Src family kinases are known to mediate microvascular endothelial hyperpermeability through phosphorylation of tyrosine residues on numerous proteins involved in endothelial cell–cell and cell–matrix adhesions (Figure 26) [201, 335, 509]. For example, Src kinase is required for the endothelial hyperpermeability in response to VEGF [122], tumor necrosis factor (TNF)-α [334], or reactive oxygen species (ROS) [229]. Src is also required to mediate the hyperpermeability induced by activated neutrophils in coronary venules [443]. In endothelial cells, Src can directly phosphorylate MLCK and increase its activity in MLC phosphorylation, leading to cytoskeletal actomyosin contraction (Figure 26) [45]. Actomyosin contraction coincides with actin stress fiber formation and endothelial hyperpermeability. At cell–cell junctions, activated neutrophils and pro-inflammatory cytokines, such as TNF-α, cause Src family kinase-dependent phosphorylation of VE–cadherin, an event believed to destabilize the adherens junction complex [202]. Within focal adhesions, both Fyn and Src associate with focal adhesion kinase (FAK) and transduce outside-in signaling triggered by integrin-ECM binding [282]. Src kinases mediate phosphorylation of FAK, and FAK binding to $\alpha_v\beta_5$ integrin, which causes endothelial hyperpermeability most likely by disrupting integrin binding to the ECM [123]. Therefore, Src alters barrier function by activating or inhibiting several intracellular signaling molecules known to affect permeability.

Src family kinases also stimulate vesicle-mediated transcytosis (Figure 26) [201, 202]. Albumin binding to the gp60 receptor triggers Src phosphorylation at Tyr-416 and activation. Once activated, Src causes phosphorylation of caveolin-1 (at Tyr-14) and dynamin-2 (at Tyr-597) and subsequent endocytosis and transcytosis of plasma components.

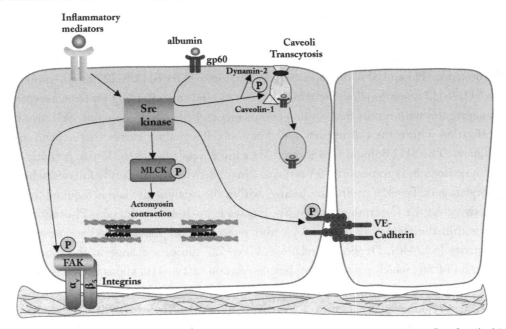

**FIGURE 26:** Src kinase-dependent mechanisms of endothelial hyperpermeability. Src family kinases (c-Src, Fyn or Yes isoforms in endothelium) phosphorylate numerous proteins in endothelial cells. The central mechanisms of endothelial hyperpermeability in response to inflammatory mediators are dependent upon Src kinase activity. Src kinases mediate gp60 receptor internalization and albumin endocytosis through Src-dependent phosphorylation of caveolin-1 and dynamin. Src kinases also phosphorylate cell junction proteins (e.g., VE–cadherin), focal adhesion kinase (FAK) at focal adhesions, and MLCK. These events cause decreased cell–cell and cell–matrix adhesion, and increased actomyosin contraction, leading to cell detachment, intercellular retraction and endothelial hyperpermeability.

## Protein Kinase C

Protein kinase C (PKC) is a family of kinases consisting of classical ($\alpha$, $\beta_I$, $\beta_{II}$, $\gamma$), novel ($\delta$, $\varepsilon$, $\eta$, $\theta$), and atypical ($\iota/\lambda$, $\zeta$) isoforms [231]. Classical (cPKC) isoforms are activated by binding to $Ca^{2+}$ and diacylglycerol (DAG); Novel (nPKC) isoforms are activated by DAG, but not $Ca^{2+}$; and atypical (aPKC) isoforms are activated by neither $Ca^{2+}$ nor DAG. PKC-$\mu$, formerly classified as an nPKC isoform, now belongs to a separate family of kinases: protein kinase D (PKD). Both phosphatidylserine and DAG bind and localize PKCs to the cell membrane. For cPKCs, $Ca^{2+}$ binding increases the affinity of PKC for membrane phosphatidylserine, and thereby increases the availability of PKC for binding to DAG at the membrane. Binding to DAG releases PKC from autoinhibition, exposing the active site of PKC for phosphorylation. Pharmacological DAG analog compounds (phorbol esters) act in a similar fashion to DAG to facilitate PKC activation. In all cases, the activity of PKC is dependent upon phosphorylation and subsequent autophosphorylation. Activated PKC, in turn, phosphorylates substrate proteins at Ser/Thr residues [11].

PKCs interact with many other molecules that are known to regulate endothelial permeability (Figure 27) [413, 509]. In the mitogen-activated protein kinase MAPK cascades, PKCs activate Ras-GTP, Raf, MAP/ERK kinases (MEKs), ERK1/2, c-Jun N-terminal kinases (JNK), and p38 MAPK. The MAPKs have been shown to play an important role in the endothelial hyperpermeability response to VEGF. Inhibition of ERK1/2 activation by the MEK1 inhibitor PD98059, or inhibition of the p38 MAPK pathway prevents VEGF- and histamine-induced albumin permeability in microvessels [492]. In other studies, the effect of PKCs in mediating endothelial hyperpermeability in response to inflammatory agents, such as thrombin and platelet activating factor (PAF) [509], is demonstrated in that treatment with pharmacological PKC inhibitors (e.g., staurosporine, bisindolylmaleimide or calphostin C) prevents the hyperpermeability-inducing effects of these agents.

The molecular targets of PKCs include endothelial cytoskeleton, junction proteins, and focal adhesion components [509]. In particular, activated PKCs affect cytoskeleton by increasing actin polymerization and activating actin binding proteins and intermediate filaments [145, 429]. PKC also induces disassembly of adherens junctions [16, 395] and causes rearrangement of focal adhesions [3]. In some cases, the effects of PKC are mediated through activation of MLCK, or inhibition of MLCP, and subsequent actomyosin contraction [413]. In microvascular endothelium, the effects of PKC on actomyosin contractility are prevented by treatment with MLCK inhibitors. Because phorbol ester or thrombin treatments increase the phosphorylation status of MLCK, and endothelial MLCK isoforms contain consensus sequences for phosphorylation by PKC within the endothelial-specific regulatory domain (a.a. 1–922) of MLCK [146, 463], it is possible that PKC acts through direct phosphorylation of MLCK. MLCK-dependent hyperpermeability can also occur through inhibition of MLCP. For example, PKC-$\alpha$ activation causes phosphorylation of GDI and release of inhibition of the small GTPase, RhoA [165, 302]; RhoA activates Rho kinase, which phosphorylates and inhibits MLCP. Thus, PKC-dependent actomyosin contractility is likely mediated through the combined effects of MLCK activation and MLCP inhibition (Figure 27).

Different PKC isoforms bind to specific scaffolding proteins: receptors for activated C kinase (RACK), which recruit activated PKC isoforms into proximity with target proteins [231]. RACK proteins direct PKCs to interact at specific locations within the cell and may account for the observation that individual PKC isoforms can play opposing regulatory roles with respect to certain cell physiological functions. For example, some PKC isoforms are known to induce endothelial hyperpermeability (PKC-$\alpha$, PKC-$\beta_{II}$, and PKC-$\delta$) [154, 460], while others (e.g., PKC-$\varepsilon$) can decrease

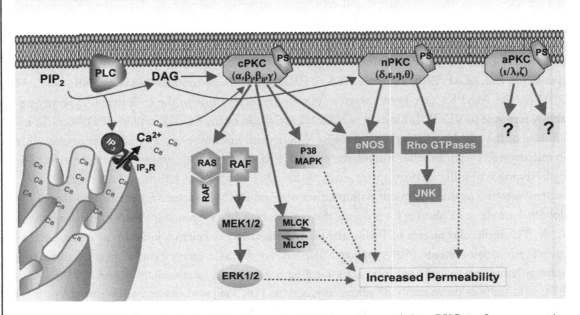

**FIGURE 27:** PKC-dependent signaling for endothelial hyperpermeability. PKC isoforms comprise classical (cPKC), novel (nPKC) and atypical (aPKC) PKCs. The cPKCs are activated by $Ca^{2+}$ and diacylglycerol (DAG); nPKCs are activated by DAG, and not $Ca^{2+}$; and aPKCs are insensitive to DAG or $Ca^{2+}$, and are activated by binding to phosphatidylserine (PS). During activation, binding to DAG or PS localizes PKC to the membrane, and PKCs are activated by autophosphorylation. Under inflammatory conditions, PKCs are activated in response to G-protein-coupled receptor activation of phospholipase C (PLC). Multiple signaling molecules are activated by PKC leading to endothelial hyperpermeability: MAP/ERK kinases, MLCK, eNOS, and small GTPases. Individual PKC isoforms have distinct roles in regulation of endothelial permeability/hyperpermeability; however, in many cases, the effects of PKC isoforms are not understood.

permeability and improve endothelial barrier function. PKC-$\alpha$ mediates endothelial hyperpermeability in response to inflammatory stimulation, as well as during hyperglycemia, ischemia, and angiogenesis [190, 509]. PKC-$\beta_{II}$ mediates hyperpermeability in the diabetic kidney and retina [154, 480]. PKC-$\delta$ mediates hyperpermeability in pulmonary endothelial cells in response to phorbol esters or DAG [442]. In the coronary system, however, PKC-$\delta$ has been shown to counteract PKC-$\alpha$ action by modulating endothelial permeability [150]. Likewise, PKC-$\epsilon$ may play a protective role in the heart and brain endothelium [315]. In coronary venule endothelial cells, PKC-$\epsilon$

causes inhibitory phosphorylation of endothelial nitric oxide synthase (eNOS), an enzyme known to mediate hyperpermeability in many types of microvascular endothelium. Other PKC isoforms phosphorylate and activate eNOS, causing increased production of NO, which has profound effects on endothelial function through NO-GC-cGMP signaling and other NO-dependent signaling events (discussed in previous sections; Figures 22 and 24) [206, 370].

## Focal Adhesion Kinase

Focal adhesion kinase (FAK) is a soluble cytosolic protein kinase that is centrally important to the organization and assembly of focal adhesions [342, 493]. The N-terminal domain of FAK contains a FERM homology region that can bind to the C-terminal domain of integrins or growth factor receptors. The C-terminus of FAK is a noncatalytically active region known as FAK-related nonkinase (FRNK) that is involved in recruitment of FAK to focal adhesions. The FRNK domain contains a FAT sequence that binds to integrins or intracellular regulatory proteins. FAK is recruited to focal adhesions at the cell membrane in response to phosphorylation. Experimentally, FRNK is often used as a natural FAK inhibitor that can block FAK recruitment or activity.

FAK contains at least five sites of tyrosine phosphorylation that are important for focal adhesion assembly and intracellular regulation [342]. Phosphorylation of Tyr-397 permits FAK binding to proteins containing the SH2 domain. In this manner, FAK recruits kinases including Src family kinases to the focal adhesion complex. Src family kinases in turn regulate the kinase activity of FAK [509]. Basal phosphorylation of FAK is important for recruiting scaffolding proteins and stabilizing focal adhesions [342, 493]. In this manner, basal FAK activity maintains normal physiological endothelial permeability. Tyrosine phosphorylation of FAK is also involved in further activation of FAK kinase activity by integrin outside-in signaling, or in response to treatment with thrombin or VEGF. Treatment with VEGF increases FAK tyrosine phosphorylation and recruitment to focal contacts.

FAK-signaled barrier responses differ according to the type of stimulus or perturbation, as well as the experimental conditions. In cultured endothelial cells, FAK kinase activity is required for enhanced barrier function under hyperosmotic conditions [369], but it contributes to hyperpermeability in the presence of thrombin [416, 457]. The latter effect is also shown in intact venules in response to histamine, phorbol esters, VEGF, or activated neutrophils [171, 494, 517]. While controversial, the general implication from studies of intact microvessels is that increased FAK activity associates with increased endothelial permeability. It has also been suggested that increased focal adhesion assembly could contribute to pathological hyperpermeability by providing a fulcrum for actomyosin contraction and the actin cytoskeleton to pull apart cell–cell junctions [509, 511]. In this way, FAK-dependent strengthening of focal adhesions would enhance endothelial barrier function under normal physiological conditions, and would facilitate barrier disruption in response to hyperpermeability-inducing agents, especially those acting through cytoskeletal contraction.

# Small GTPases

Guanosine triphosphatases (GTPases) are signaling molecules that control the activity of multiple enzymes including protein kinases. GTPases are normally bound to GDP in the inactive state and become activated when GDP is replaced by GTP (Figure 28) [427]. The exchange of GDP for GTP is catalyzed by guanine exchange factors (GEFs). GTPase activation in turn depends upon conformational changes induced by GTP binding which allow GTPases to bind and activate target enzymes. Negative regulation is performed by GTPase-activating proteins (GAPs) that induce dissociation of GTP, and by guanine nucleotide-dissociation inhibitors (GDIs), enzymes that retain GTPases in the GDP-bound state.

Small GTPases are 21–25 kDa signaling molecules of the Ras superfamily that include the Rho, Ras, Rab, Ran and Arf families of proteins [427]. The Rho family contains several members that are important for regulation of cytoskeleton, adhesion, and endothelial barrier function. Rho family GTPases may induce hyperpermeability (e.g., RhoA) (discussed below) or protect against endothelial barrier dysfunction (e.g., Rac-1, Cdc42) (discussed in the next chapter).

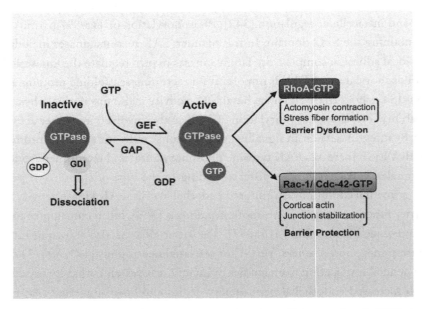

**FIGURE 28:** Small GTPases in control of endothelial barrier function. Small GTPases play central roles in endothelial barrier dysfunction (e.g., RhoA) and barrier protection (e.g., Rac-1 or Cdc-42). Small GTPases are bound to GDP in the inactive state, stabilized by GDP-dissociation inhibitors (GDIs). Upon release from GDI binding, GDP can be exchanged for GTP by GTP-exchange factors (GEFs). Small GTPases become active once in the GTP-bound state. Inactivation occurs when GTP is exchanged for GDP by GTPase-activating proteins (GAPs).

Activation of RhoA increases microvascular endothelial permeability [427, 460], whereas basal RhoA activity maintains normal physiological barrier function in microvascular endothelium, i.e., basal permeability is increased upon exposure to RhoA inhibitors. RhoA is also centrally involved in endothelial hyperpermeability through effects on multiple cellular events including actin stress fiber formation and increased actomyosin contractility. RhoA may become activated through activation of RhoGEF or through inhibition of RhoGDI. For example, thrombin exposure induces endothelial hyperpermeability, mediated through activation of RhoA. Hyperpermeability in response to thrombin is mediated through activation of Rho kinase (ROCK), which in turn phosphorylates and inactivates MLCP. MLCP inhibition increases MLCK-dependent phosphorylation of MLC and actomyosin contractility. Actomyosin contraction causes retraction at cell junctions, and is accompanied by other molecular events that contribute to endothelial hyperpermeability. In brain endothelium, RhoA-dependent hyperpermeability is accompanied by phosphorylation of occludin, which is believed to destabilize tight junctions [194, 460]. In many types of microvascular endothelium, RhoA activation induces actin stress fiber formation.

In contrast to RhoA, two other small GTPases, Rac-1 and Cdc42, decrease endothelial permeability and improve barrier function [247, 460, 473]. These small GTPases are activated downstream of barrier protective signaling in response to sphingosine-1-phosphate (S1P), or during resolution (late stages) of hyperpermeability caused by inflammatory agents. Therefore, activation of Cdc42 and Rac-1 is often considered a feedback/repair mechanism following endothelial barrier dysfunction. The molecular mechanisms of Rac-1 and Cdc42 mediated protection are not fully understood; however, these molecules cause rearrangements of the cortical actin cytoskeleton. Rac-1 overexpression induces F-actin polymerization at the periphery (cortical actin) of endothelial cells [473]. F-actin polymerization may be due in part to inhibition of the actin-severing enzyme cofilin. Following S1P treatment, Rac-1 binds to p21-associated kinase (PAK) and activates LIM-1; LIM-1 in turn inactivates cofilin and prevents actin-severing activity. Rac-1 also increases cortical F-actin polymerization by inducing cortactin translocation to the cell cortex. In addition, both Rac-1 and Cdc42 stabilize cell–cell junctions by preventing dissociation of α-catenin from the adherens junction complex [309, 460].

# PROTEASES AND THE EXTRACELLULAR MATRIX

The extracellular matrix (ECM) participates in endothelial barrier responses by way of binding to endothelial cell surface integrins at focal adhesions. The ECM also serves as a reservoir for extracellular signaling molecules. Important permeability-regulating factors, including RGD-containing matrix proteins (e.g., fibrinogen, fibronectin), are embedded in the ECM, and they are released after degradation by proteases and other enzymes [99, 170, 373]. In some cases, depletion of ECM components increases endothelial permeability, as is the case for hyaluronan degradation by hyaluronidase. In other instances, as is the case for vascular endothelial growth factor (VEGF), enzymatic degradation of ECM releases matrix-bound growth factors that can bind to endothelial cell surface receptors and induce hyperpermeability. Hence endothelial barrier function is affected by degradation both directly by weakening the integrity of the ECM, and indirectly by releasing signaling molecules that bind to endothelial cell surface receptors. Another mechanism of signaling by extracellular proteases is through cleavage of membrane protein extracellular domains (shedding), as in the case of syndecan shedding from the glycocalyx [292]. Proteolytic cleavage releases soluble glycoprotein ectodomains into the ECM or into the blood circulation to function as receptor ligands or soluble receptors. Two major families of enzymes have been implicated in matrix degradation and membrane receptor shedding: matrix metalloproteinases (MMPs) and a disintegrin and metalloproteinases (ADAMs).

## MMPs

Matrix metalloproteinases are a family of 29 proteases that are dependent upon metal (zinc) ions for their proteolytic activity [133, 412]. MMPs consist of collagenases, gelatinases, stromelysins, matrilysins, and membrane-type MMPs (MT-MMPs) [133, 324, 468]. The collagenases (MMP-1, -8, and -13) cleave collagen and other substrates, e.g., MMP-1 also cleaves and activates the protease-activated receptor (PAR-1), similar to the hyperpermeability-inducing agent thrombin [53]. Gelatinases (MMP-2, and -9) digest gelatins and collagens [324]. Stromelysins (MMP-3, -10 and -11), and matrilysins (MMP-7 and -26) digest various components of the ECM other than collagen. Matrilysins differ from other MMPs in that they lack a homopexin domain. Membrane-type MMPs (MT-MMPs) are, as the name implies, bound to the plasma membrane, either as integral membrane proteins (MMP-14, -15, -16, and -24) or via a glycosylphosphatidylinositol (GPI) anchor (MMP-17 and -25). MT-MMPs have widely varying substrate specificities and tissue distributions. The remaining MMPs are grouped together as "Other MMPs," each exhibiting different specialized activity (MMP-12, -19, -20, -23, -27, and -28).

MMPs are characterized by a catalytic protease domain with the zinc binding sequence: HEXGHXXGXXH and a prodomain with a cysteine switch motif (PRCGXPD) (Figure 29, left) [324, 468]. Inactive precursor pro-MMP is maintained in the inactive state while the cysteine (C) residue of the switch motif is bound to the zinc metal ion positioned between the three histidine (H) residues of the catalytic domain [431]. MMPs are activated by cleavage of the prodomain of pro-MMPs, thereby removing the cysteine switch motif and unmasking the zinc metalloprotease domain [324]. Most MMPs also contain a homopexin-like domain that confers substrate specificity and may facilitate binding to other proteins (Figure 29, left). Specialized functions of MMPs are dictated by structural domain variations. For example, membrane-type (MT)-MMPs have an

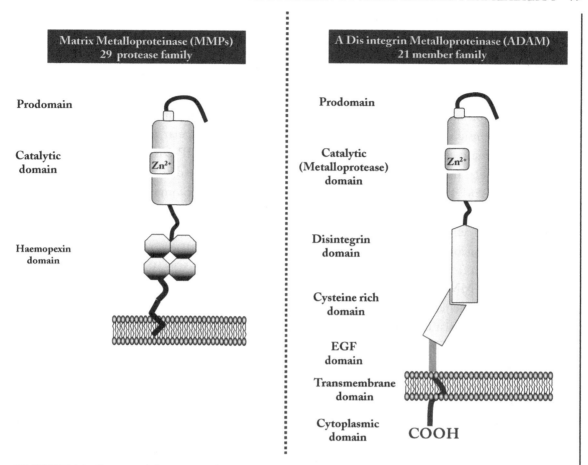

**FIGURE 29:** Structural features and similarities of MMPs and ADAMs. MMP and ADAM family members are cell surface-expressed integral membrane proteins, with extracellular protease activity conferred by a conserved zinc-binding catalytic metalloproteinase domain. Metalloproteinase activity is activated by cleavage of the adjacent autoinhibitory prodomain. MMPs are anchored to the plasma membrane by a membrane-bound C-terminal domain; Some MMPs have a hemopexin domain that confers substrate specificity. In contrast to MMPs, ADAMs are transmembrane-spanning proteins with a cytoplasmic C-terminal domain that is involved in intracellular signaling. ADAMs also possess an extracellular disintegrin domain for binding to integrins, as well as an epidermal growth factor (EGF) domain, and a cysteine-rich domain.

additional stretch of 70 to 100 hydrophobic C-terminal amino acids that inserts into and anchors these MMPs to the membrane.

Activated MMPs cleave components of the endothelial extracellular matrix during normal remodeling and maintenance of the basement membrane and can cause microvascular barrier dysfunction under pathophysiological conditions [133]. MMPs play a major role in remodeling the ECM during formation of new blood vessels (angiogenesis) [133]. For this reason, MMP-2 and MMP-9 are considered necessary for wound healing or tumor vascularization. Consistent with a role in angiogenesis and wound healing, elevated vascular tissue MMP-2 is a significant predictor of successful vascular remodeling and maturation of surgically implanted arteriovenous fistulas in human patients [258]. Therefore, MMP-2 may be a useful molecule for promoting vascular remodeling.

MMPs affect multiple physiological or pathophysiological processes by cleaving membrane proteins on the cell surface. For example, MMPs cleave and activate membrane-bound growth factors (e.g., transforming growth factor beta (TGF-β), epidermal growth factor (EGF)), as well as release growth factors that are embedded within the ECM (e.g., VEGF, basic fibroblast growth factor (bFGF)). MMP-2 also cleaves the tight junction proteins occludin and claudin-5 and contributes to brain edema during ischemia–reperfusion injury [384]. Blood cells (leukocytes) also express MMPs that assist in their extravasation from the blood during tissue inflammation. MMP-9 secretion by neutrophils contributes to endothelial hyperpermeability by cleaving components of the extracellular matrix (ECM) (e.g., collagen), by cleaving and activating cytokines (e.g., CXCL5, CXCL6, IL-8), and by releasing VEGF from the ECM [189].

MMPs are inhibited by tissue inhibitors of matrix metalloproteinases (TIMPs) [133, 324]. There are four members of the TIMP family of proteins (TIMP1–4) [324]. TIMP-1, -2, and -4 are soluble, secreted proteins, whereas TIMP-3 is membrane-bound [257]. TIMPs have an N-terminal domain that binds to MMPs and inhibits MMP activity [322]. A pair of cysteine residues forms a disulfide bridge within the N-terminal domain that inserts into the MMP catalytic domain and chelates the Zinc ion [163]. There is great interest in developing compounds based on this structure for therapeutic inhibition of MMP because of the involvement of MMPs in many pathophysiological processes. For example, MMP-2 is involved in tumor angiogenesis and metastasis [124, 295, 404], promoting both tumor cell migration across the vascular endothelium during intravasation from the primary tumor into the circulation and extravasation from the blood during invasion of remote tissues [87, 116]. MMPs are expressed by invasive tumor cells and by vascular endothelium, suggesting that MMP-2 at either location could be involved in metastatic tumor cell migration across the vascular endothelium. It was recently shown that transendothelial migration of metastatic breast cancer requires a complex interaction of TIMP-2 on the tumor cell, with MMP-2 and MMP-14 on lung microvascular endothelial cells [412]. In response to breast cancer cell attachment, lung microvascular endothelial cells produce active MMP-2 [222]. Furthermore, gene knockdown of MMP-2 expression in the endothelial cells prevented tumor cell transmigration. MMP-14, a proteolytic activator of MMP-2, is localized to the endothelial cell surface [412], and MMP-2 is activated by forming a complex with endothelial MMP-14 together with TIMP-2 from the tumor cells [214, 248, 433]. This regulatory complex mediates endothelial barrier dysfunction (decreased TER) in response to tumor cells binding to the endothelium [412].

## ADAMs

ADAMs (a disintegrin and metalloproteinases) are a family of 40 known genes, encoding at least 21 human functional proteins involved in tumor metastasis, angiogenesis, and inflammation [50, 75, 117]. ADAMs are type 1 integral membrane glycoproteins, possessing cytosolic, transmembrane, and extracellular structural regions, with 5 distinct extracellular domains (Figure 29, right). Similar to MMPs, ADAMs have an extracellular N-terminal prodomain, and a catalytic zinc-binding protease domain, most frequently with the amino acid consensus sequence HEXGHXXGXXHD. Metalloproteinase activity is suppressed by the prodomain and is induced by convertase-catalyzed cleavage and removal of the prodomain [75]. ADAM metalloproteinase activity selectively cleaves components of the ECM, including collagen and gelatin, in an isoform-specific manner [294]. For some ADAM isoforms, protease activity is used to cleave the extracellular domains of other membrane proteins, including cadherins [327, 375, 401] or growth factor receptors [293]. Cleavage products released by this process (shedding) can act as extracellular signaling molecules, as occurs for production of epidermal growth factor (EGF) and other endogenous ligands of the EGF receptor (EGFR) [387]. Ectodomain shedding can trigger additional cleavage of the intracellular domain and/or liberation of the soluble remainder of the membrane protein into the cell interior [117]. This process, known as regulated intracellular proteolysis (RIP), permits further signaling by the soluble intracellular fragment, including nuclear regulation of gene transcription.

Many ADAM family members have additional extracellular domains: a cysteine-rich region, an EGF-like domain, and a disintegrin domain, often with the consensus sequence $RX_6DLPEF$, for binding to integrins [117]. ADAMs binding to integrins may disrupt endothelial adhesion to the ECM during tumor or leukocyte extravasation during inflammation and angiogenesis [75]. One unique isoform, ADAM15, has an additional RGD sequence for binding to integrins [329] that is important for cell–cell adhesion between leukocytes and endothelium during inflammation [75]. During inflammation, ADAM15 mediates endothelial hyperpermeability via a mechanism that does not require the proteolytic activity of its extracellular domain [435]. Rather, the cytosolic C-terminus initiates signaling through Src kinase and ERK1/2 to signal endothelial hyperpermeability. This signaling mechanism contributes to microvascular fluid leakage in the lungs of mice in response to systemic exposure to bacterial endotoxin (lipopolysaccharide), and is involved in pneumonia-induced lung injury in humans (unpublished observations by S.C. Sun and S.Y. Yuan).

Therefore, signaling through ADAM15 and other ADAMs is multifaceted, involving multiple functions of extracellular and intracellular domains of ADAM proteins.

## Fibrinogen

Fibrinogen is a plasma protein and an integral component of the ECM under normal physiological conditions. This molecule is composed of two sets of polypeptide α, β, and γ chains forming an elongated structure with two outer globular D-domains that are connected to central E-domains through coiled-coil segments [113, 281, 318, 501]. In the circulation, fibrinogen plays an essential role in coagulation. In response to activation of intrinsic or extrinsic coagulation factors, thrombin cleaves fibrinogen, converting it to fibrin, which forms cross-linked fibrin networks, i.e., blood clots. Coagulation is followed by fibrinolysis, plasmin-mediated proteolysis of fibrin that disassembles and removes the blood clot, and in doing so generates fibrin degradation products (FDPs) [72, 143]. Typical FDPs include D-dimers and monomers containing the C-termini of the polypeptide chains that are measured as clinical markers of thrombotic disease [6, 198, 472, 519].

Fibrinogen serves as an important structural component in the ECM that retains water, and provides scaffolding for numerous proteins and growth factors [134, 425]. Because fibrinogen is a plasma protein, its presence in the ECM may be due to deposition from the circulation. Under injurious conditions, an important source of FDPs is the degradation of fibrinogen and fibrin from the extravascular matrix of injured or inflamed tissues [93, 254, 296, 357]. The circulating level of FDPs has been correlated with the severity of systemic inflammatory response syndrome (SIRS) in trauma patients and thus used as a marker of systemic injury and multiple organ dysfunction [148, 209, 254, 296, 470].

Fibrinogen exerts pathophysiological impact well beyond its role in hemostasis. Elevated plasma levels of fibrinogen are associated with atherosclerosis, myocardium infarction, and peripheral vascular disease [6, 472, 519]. However, it remains controversial as to whether plasma fibrinogen is merely an acute phase protein produced during inflammation or is the cause of inflammation [227]. Likewise, effects of fibrinogen on vascular permeability have been reported [18, 152, 385, 450], but with no consensus regarding whether fibrinogen as an intact (340 kDa) protein disrupts or protects barrier function. Investigation is often confounded by the multi-domain structure of fibrinogen and its diverse targets, ranging from platelets and leukocytes to endothelial cells [6]. In general, fibrinogen per se is considered to have a minimal effect on permeability due to its unique structure: many reactive regions are buried in loops or pockets and are only active in certain structural conformations [113, 254, 281, 318, 502]. Proteolytic digestion or molecular truncation that exposes reactive regions alters the cellular targets and biological functions of fibrinogen. For example, fragments with exposed endothelial binding sites are more potent than native proteins in regulating endothelial functions. This may explain the ineffectiveness of Ancrod, a fibrinogenolytic agent intended to treat myocardial ischemia–reperfusion by reducing the plasma level of fibrinogen. In cleaving fibrinogen, Ancrod also generates FDPs that have varied effects on vascular function [519].

Because of the lack of therapeutic benefit from strategies aimed at decreasing fibrinogen, investigators have directed their research toward small FDPs, rather than the native form of fibrinogen. An array of peptides corresponding to various segments of fibrinogen subunits has been developed and characterized for their biochemical properties [12, 503, 504]. Among them, the 30-kDa C-terminal fragment of fibrinogen-$\gamma$ ($\gamma$C) has been identified as a potent inducer of cell apoptosis and inhibitor of capillary tube formation [12, 503, 504]. Further studies demonstrate a new function of this FPD in causing microvascular leakage by binding to the endothelium. Several sequences in $\gamma$C are suggested to mediate its binding to endothelial or circulating cells, including $\gamma$117–133 that binds ICAM-1 [396], $\gamma$377–395 that binds CD11b/CD18 [139, 319, 451], and $\gamma$400–411 that binds the $\alpha_{II}\beta_3$ integrin [39, 129]. Interestingly, a recent study shows that $\gamma$C-induced plasma leak in mesenteric microvessels is greatly attenuated in $\beta_3$ integrin knockout mice [170], supporting the importance of this integrin in mediating endothelial barrier dysfunction caused by FDPs.

.   .   .   .

CHAPTER 6

# Endothelial Barrier Protectors

## PHOSPHOLIPIDS

Blood cells, especially platelets, contain phospholipids that protect endothelial barriers from hyperpermeability [309]. The protective effects of platelets were originally noted in several experimental animal models where infusion of platelet-rich plasma was shown to protect vascular tissues from plasma leakage [347]. It was initially suspected that barrier protection was due to platelets directly plugging the sites of leakage [216]. However, it was later shown that treatment with either platelet-rich or platelet-conditioned medium (PCM) (in the absence of intact platelets) is sufficient to block endothelial leakage in response to hyperpermeability-inducing agents [180, 279]. Initial biochemical analysis of PCM revealed the barrier-protective agent lysophosphatidic acid (LPA), a membrane phospholipid [17]. However, the barrier-protective properties of LPA [310] are controversial because exposure to LPA increases permeability in brain capillary endothelial cells [402, 459]. Therefore it appears that the protective effects of LPA on microvascular barriers are tissue-specific.

### Sphingosine-1-Phosphate

More recently, it was shown that another platelet-derived phospholipid, sphingosine-1-phosphate (S1P), has barrier-protective properties [147]. S1P is a much more potent and broadly acting barrier protective agent than LPA [309]. Normally, S1P is maintained at low levels in the blood ($\leq 1.1$ $\mu$M) due to the activity of sphingosine lyase (degrades S1P) and sphingosine phosphatase (dephosphorylates S1P) [301, 366, 473]. However, large amounts of S1P are generated by sphingosine kinase and are stored in platelets. During trauma or inflammation, platelets release S1P into the blood [309].

The actions of S1P on endothelial barriers are mediated by binding to S1P receptor isoforms 1–5 (S1P$_{1-5}$), GPCRs on the cell surface (Figure 30) [473]. At high concentrations (>5 $\mu$M) S1P binds to lower affinity S1P receptors (S1P$_{2/3}$) and has negative effects on barrier function, including actin stress fiber formation through activation of G$\alpha_{12/13}$ or G$\alpha_q$, and Rho GTPase [301, 473]. At low concentrations (<1 $\mu$M), S1P binding to S1P$_1$ on endothelial cells mediates barrier-protective effects by activating the pertussis toxin-sensitive G$\alpha_i$ protein, followed by activation of Rac-1 (Figure 30). The requirement for S1P$_1$ in barrier protection is supported by studies of transgenic SIP$_1$

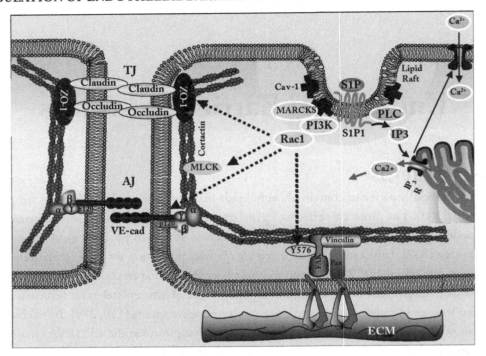

**FIGURE 30:** Sphingosine-1-phosphate (S1P) signaling for endothelial barrier protection. S1P mediates barrier protection through binding to the $S1P_1$ receptor. $S1P_1$, a G-protein-coupled receptor localized in caveolin (cav-1)-rich lipid raft microdomains (caveolae), recruits signaling molecules (MARCKS, PI3K) and activates PLC, triggering $IP_3$-dependent $Ca^{2+}$-induced $Ca^{2+}$ influx, and activating multiple intracellular signaling molecules, including the small GTPase Rac-1. Rac-1 signaling strengthens cell–cell junction and focal adhesion complexes, and relocates MLCK and cortactin to the cell peripheral cortical actin band. Elevated $Ca^{2+}$ and other signaling events increase actomyosin contractility at the cell periphery. Combined effects on cell junctions, focal adhesions and cell morphology decrease endothelial permeability and increase barrier integrity.

knockout mice, where S1P fails to protect against microvascular leakage in the lungs in response to the bacterial endotoxin lipopolysaccharide (LPS) [390].

The mechanisms of S1P-mediated barrier protection are not clearly understood. S1P signaling in endothelium is exceptionally complex and is summarized well in a recent review [473]. S1P affects cell–cell junctions, actin cytoskeleton and focal adhesions through signaling initiated in caveolae (caveolin-rich lipid rafts) (Figure 30). Exposure of endothelial cells to S1P induces recruitment of $S1P_1$ to lipid rafts, followed by threonine phosphorylation of $S1P_1$. $S1P_1$ activation causes

IP$_3$-dependent release of intracellular Ca$^{2+}$ and recruits several signaling molecules, including myristoylated alanine-rich protein kinase C substrate (MARCKS) and phosphatidylinositol-3-kinase (PI3K), resulting in Rac-1 activation. S1P$_1$ activation stabilizes endothelial cell–cell junctions by increasing VE–cadherin and β-catenin localization at adherens junctions and increasing ZO-1 abundance at tight junctions. S1P$_1$ activation also induces phosphorylation of FAK (at Y576) and FAK-dependent reorganization of focal adhesions at the cell periphery. However, it is not clear how focal adhesion reorganization contributes to barrier stabilization in response to S1P treatment.

Low concentrations of S1P increase phosphorylation of MLC and modestly increase cytoskeletal tension [309, 473]. This is in contrast to the barrier-protective signaling molecule cAMP, which decreases MLC phosphorylation and relaxes the cytoskeleton. Yet both cAMP and S1P increase cell spreading, increasing cell–cell contact between juxtaposed endothelial cell margins and decreasing permeability. Barrier protection by increased MLC phosphorylation and actomyosin contractility is paradoxical in view of the well-accepted notion that these actions cause hyperpermeability, but may be explained by differences in the subcellular localization of activated MLCK [413, 473]. For example, treatment of lung endothelial cells with S1P induces lipid raft formation and re-localization of activated MLCK to raft domains (caveolae) at the cell membrane [520]. Interestingly, the raft-localized MLCK is predominantly endothelial isoforms MLCK1 and MLCK2. Within raft domains, MLCK binds to cortical actin [115], associates with activated tyrosine kinases [520], and is phosphorylated by Src family kinases. Hence, these observations suggest that modestly increased actomyosin-based tension in cortical actin filaments (in the absence of stress fibers) improves the efficiency of cell–cell contacts and decreases endothelial permeability. This concept, as well as the defined roles of individual MLCK isoforms, remains to be further elaborated.

## ACTIVATED PROTEIN C

Protein C (factor XIV) is a serine protease present in the blood circulation that plays a role in the maintenance of hemostasis [134, 273, 454]. Protein C is a proenzyme member of the coagulation system that is cleaved by thrombin into an active anticoagulant, activated protein C (APC). During coagulation, protein C becomes bound to the endothelial cell protein C receptor (EPCR) and presented to membrane-bound thrombomodulin [127], a receptor present on the surface of vascular endothelial cells that complexes with thrombin and catalyzes conversion of protein C to APC.

APC is known to exert a protective effect in the pulmonary microvasculature during sepsis and acute lung injury (ALI) [134]. APC also prevents endothelial hyperpermeability in response to direct application of thrombin [135]. This permeability-reducing effect may be mediated through activation of sphingosine kinase, production of S1P, and activation of Rac-1. Furthermore, APC binding to EPCR facilitates S1P signaling by co-localizing with the receptor S1P$_1$. This association is negatively regulated by thrombin signaling through ADAM17, which causes cleavage and shedding of EPCR. Hence, thrombin and APC are opposing regulators of coagulation and of endothelial hyperpermeability.

Several anti-inflammatory effects of APC have been attributed to the protective effects of APC in vivo. For example, APC inhibits nuclear factor kappa-B (NF-κB) and decreases cytokine production in leukocytes following exposure to LPS [218, 219]. APS also inhibits thrombin production during coagulation by inhibiting coagulation factors Va and VIIIa [454]. This is suggestive of a mechanism by which APC may decrease hyperpermeability during inflammation.

# ANGIOPOIETINS

Angiopoietins (Ang 1–4) are structurally related, soluble secreted proteins that affect endothelial cell function [28], of which Ang-1 and -2 are potent modulators of endothelial permeability and barrier function. Endothelial cells secrete Ang-2 for autocrine signaling [132], whereas Ang-1 is secreted by parenchymal cells of the extravascular tissue (i.e., parietal cells, smooth muscle cells, and fibroblasts) onto endothelial cells for paracrine signaling [28, 104]. In addition, Ang-1 binds to ECM proteins and can be stored in the ECM [499], whereas Ang-2 cannot bind to the ECM and is stored within endothelial cells [132].

Ang-1 and Ang-2 bind to endothelial Tyrosine kinase with Immunoglobin (Ig) and Epidermal Growth Factor (EGF)-like homology domain receptors (TIE-1 and TIE-2) [28, 213]. As the acronym implies, TIEs are integral membrane receptor tyrosine kinases with extracellular Ig-like and EGF-like domains. TIEs also have an extracellular fibronectin domain and an angiopoietin-binding domain. Binding of Ang-1 or Ang-2 to TIE receptors are mediated by a fibrinogen-like domain. Ang binding induces tyrosine kinase phosphorylation and triggers multiple intracellular signaling pathways. Similar signaling occurs through either TIE-1 or TIE-2 [240], but TIE-2 is much more important for endothelial physiology [28].

Ang-1 binding to TIE-2 is barrier protective, decreasing resting permeability and maintaining endothelial cells in a quiescent state [28, 280]. Ang-1 expression is also pro-angiogenic and increases during neovascularization. In contrast, Ang-2 is a receptor antagonist of Ang-1 [131]; Ang-2 also activates TIE-2 signaling, though at very high unphysiological concentrations. Elevated Ang-2 in the blood circulation occurs in sepsis, trauma and inflammatory diseases [28], and is correlated with pulmonary vascular hyperpermeability and increased mortality in patients with sepsis and acute lung injury [341, 455]. This suggests that the in vivo role of Ang-2 is to compete with the protective effects of Ang-1 and exacerbate pathophysiological conditions. A more recent finding is that soluble (s) Tie-2 receptor is increased in the circulation of sepsis patients [456]. Elevated sTIE-2 concentrations result from cleavage and shedding of the extracellular domain of TIE-2 by MMPs in response to VEGF exposure. Circulating sTIE-2 may prevent the protective effects of Ang-1, by binding to Ang-1 and reducing the amount of Ang-1 available for binding to endothelial TIE-2.

These observations support the conclusion that Ang-1/TIE-2 signaling is important both for protection of endothelial barriers, and for maintenance of normal physiological barrier function.

Activation of TIE-2 by Ang-1 binding induces tyrosine kinase activation and recruitment of intracellular signaling molecules, including the p85 subunit of PI3K [28, 213, 280]. PI3K activates AKT signaling, which has multiple effects on intracellular regulation. For example, AKT activation causes transcriptional down-regulation of Ang-2 production in endothelial cells [101], which may suppress Ang-2 antagonism of Ang-1. TIE-2 activation by Ang-1 is shown to prevent the hyperpermeability-inducing effects of platelet-activating factor, bradykinin, histamine [354], or VEGF [151]. Protection against VEGF-induced permeability by Ang-1/TIE-2 is mediated through sequestration of soluble Src kinase, a central mediator of the VEGF-induced hyperpermeability response. Ang-1 treatment also prevents VEGF-induced TRPC1 $Ca^{2+}$ channel activation and $Ca^{2+}$-dependent hyperpermeability [215].

Additional endothelial-protective effects of Ang-1/TIE-2 signaling include increased production of S1P, through activation of ERK1/2 and sphingosine kinase (protective effects of S1P are discussed above) [272]. TIE-2 activation by Ang-1 additionally inhibits nuclear factor kappa-B (NF-κB), a transcription factor that mediates production of inflammatory cytokines and expression of cell adhesion molecules that facilitate leukocyte binding to the endothelium during inflammation [213]. Hence, multiple mechanisms mediate the barrier-protective effects of Ang-1/TIE-2 signaling in endothelium.

CHAPTER 7

# Pathophysiology and Clinical Relevance

## BURN EDEMA

Thermal injury affects more than 2 million Americans each year. Clinically, burns are evaluated based on wound depth (degree) and coverage area (% of total body surface area, TBSA). A first-degree burn causes redness and swelling in the outmost layers of skin (epidermis). A second-degree burn involves redness, swelling, and blistering with damage extended beneath the epidermis to deeper layers of skin (dermis). A third-degree burn, also called a full-thickness burn, destroys the entire depth of skin and may extend to the underlying fat, muscle, or bone, causing significant scarring. Major (or severe) burns are defined as first- or second-degree burns covering >25% TBSA in adults (>20% in children), or a third-degree burn covering >10% BSA.

Despite the remarkable improvement in critical care and wound management, systemic complications remain a major cause of mortality and morbidity in patients with severe burns even after successful initial resuscitation. Of great concern is the development of massive edema accompanied by a systemic inflammatory response syndrome (SIRS) that affects multiple organs and tissues. As a cardinal component of systemic inflammation, microvascular leak occurs not only locally at the burn wound, but also in distant tissues remote from the wound [27]. The leak response peaks within the first 1–6 hours and starts to resolve by 6–12 hours [58, 70, 109, 284, 352]. Loss of plasma fluid is the major cause of circulatory (hypovolemic) shock. Accumulation of protein-rich fluid in tissues impedes the local microcirculation and blood–tissue exchange, resulting in tissue malperfusion and hypoxia [8, 105, 109, 196, 434]. Intensive fluid therapy helps maintain the circulatory stability of patients during the resuscitation period; however, without effective approaches to stop vascular leakage, excessive fluid administration exacerbates edema contributing to adult respiratory distress syndrome (ARDS), abdominal compartment syndrome, and, in worse cases, multiple organ failure [27, 105, 169, 284].

Hypotheses that explain the onset and progress of plasma accumulation in tissues include increased fluid filtration pressure (due to vasodilation), blockage of lymphatic clearance, and increased interstitial osmotic activity [26, 109, 236, 284, 352]. With respect to the latter, it is believed that the thermal effect on tissues causes breakdown of the scaffolding matrix (collagen, collagen, hyaluronic acid, etc.), leading to increased interstitial compliance and accumulation of matrix fragments in the extravascular space. These small fragments serve as osmotically active molecules that produce a "sucking" force for the outward flux of plasma fluid. Based on this theory, clinicians formulate fluid therapies for burn patients centering on volume supplementation while correcting the imbalance of the Starling forces. A variety of resuscitation fluids have been developed, including hypertonic saline and fluids containing colloids or macromolecules (albumin, dextran, and PEG). Although these solutions seem to be highly effective in ameliorating or delaying the progress of burn edema in animals, their beneficial effects and efficacy in human patients are not definitive. Currently, hypertonic saline and colloid fluids are under extensive clinical evaluation for long-term outcomes in trauma patients and for identifying their mechanisms of action at the cellular and molecular level.

Microvascular hyperpermeability has thus far been considered the most important mechanism of edema, especially in tissues remote from the burn wound site [91, 142, 410]. Accordingly, an array of permeability-increasing mediators, including histamine, prostaglandins, and cytokines, has been identified in the circulation of burn patients [26, 27, 158]. In particular, tissue injury-coupled activation of complements triggers mast cell degranulation. Mast cells are located in proximity to blood vessels in all tissues. As the major granule content of mast cells, histamine is released and acts on the endothelium, causing junction opening and increased paracellular permeability (molecular details described in the previous chapters). Meanwhile, multiple cascade systems in human plasma can be triggered by thermal injury, including the arachidonic acid cascades (producing prostaglandins), the kallikrein cascade (producing bradykinin and plasmin), the complement cascades (activating C3 and C5a), and the coagulation/fibrinolytic cascades (forming fibrin clots and fibrin degradation products). The majority of these products are capable of increasing microvascular permeability. In addition, cytotoxic metabolites and apoptotic mediators produced by activated leukocytes and macrophages (such as oxidants and proteases) contribute to the injurious process [398, 476, 505]. More details regarding the role of leukocytes in microvascular barrier injury are provided in the following section (ischemia–reperfusion injury).

However, despite the progress in identifying the biological sources and chemical natures of edematous factors, attempts to block individual mediators as therapeutic means for treating burn edema have met with limited success. For example, a clinical trial with leukocyte antiadhesion therapy was discontinued due to high mortality [90, 157]. Histamine antagonists have been used in burn patients but their therapeutic significance is questioned along with concerns about their adverse hemodynamic effects [27]. Prostaglandin inhibitors fail to improve microvascular function while delaying wound healing and causing perioperative bleeding [27, 83, 323], and COX-2 inhibitors cause cardiovascular distablization while showing minimal protection in immunocompetent subjects [136, 323]. On the other hand, steroids are efficacious as anti-inflammatory therapeutics, but are used restrictively as they suppress the host defense and increase the risk of infection and bone degeneration [27]. Overall, many of the anti-inflammatory drugs exert unwanted effects, and the complex interactions among multiple inflammatory pathways underscore the limited effectiveness of individual antagonists [83, 157, 168, 323]. These problems signal the need for development of alternative clinical strategies. Interventions or therapeutic agents directly targeting the end-point cellular processes that ultimately open the microvascular barrier may lead to a new direction in trauma research.

MLCK is a common endpoint effector for multiple signaling pathways triggered by circulating inflammatory factors in response to trauma. Recent molecular studies in animal models of burn injury support the therapeutic potential of MLCK inhibition in attenuating microvascular leakage. Treatment with ML-7 significantly attenuates burn-induced splanchnic microvascular hyperpermeability in a dose-dependent manner [205]. The construction of MLCK-210 knockout mice [471] has enabled in vivo testing of the nmMLCK-mediated microvascular barrier response to stress or injury [285]. Compared with wild-type mice which show increased albumin transflux and hydraulic conductivity in intestinal and mesenteric microvessels after severe burns, MLCK-210 knockout mice demonstrate attenuated microvascular leak and improved survival after severe burns [381]. In contrast, endothelial-specific overexpression of MLCK2 in transgenic mice shows enhanced plasma protein leakage into lung tissue during sepsis or injury, suggesting that individual nmMLCK isoforms mediate endothelial hyperpermeability in vivo [313]. Currently, research efforts are being devoted to the development and testing of MLCK isoform-specific inhibitors as potential therapies for microvascular injury during trauma or sepsis.

## ALI/ARDS AND SEPSIS

Acute lung injury/acute respiratory distress syndrome (ALI/ARDS) is a severe form of pulmonary dysfunction characterized by hypoxemia and pulmonary edema in the absence of congestive heart disease. Diffuse inflammation, leukocyte infiltration, and intra-alveolar formation of protein-rich hyaline membranes on denuded alveolar epithelial surface represent the major pathological features of the syndrome. Its etiology includes direct injury (inhalation of noxious gases, aspiration of gastric contents, or infection) and secondary effects of systemic inflammation in response to trauma, sepsis, thrombosis, or acute pancreatitis. These processes damage both the microvascular endothelium and alveolar epithelium that constitute the alveolar capillary membrane barrier. Failure of this barrier is the most important cause of plasma leakage into alveoli [437, 477]. Fluid in the alveoli decreases local blood flow and surfactant production, leading to reduced lung compliance with impaired oxygen diffusion capacity. As inflammation progresses, pulmonary fibrosis develops and gas exchange is further impeded.

Complex mechanisms are involved in alveolar capillary membrane hyperpermeability. Multiple inflammatory agents, released from activated leukocytes or endothelial cells in response to systemic inflammation, are considered pathogenic factors for ARDS, including TNF-α, IL-1, and

IL-6 [42, 203]. The level of TNF-α in the serum and bronchoalveolar lavage fluid (BALF) is substantially increased in ARDS patients [36, 57], and blocking TNF-α ameliorates pulmonary edema in animal models of ARDS [175]. The mechanisms by which TNF-α causes barrier injury include actomyosin contraction-driven intercellular gap formation, as well as NF-κB-dependent upregulation of pro-inflammatory cytokines (e.g., IL-1β) and adhesion molecules (e.g., ICAM-1) [49, 379]. Within this context, IL-1β is another pro-inflammatory cytokine found in BALF from ARDS patients, though its detrimental role in the disease is challenged by an in vitro study showing that IL-1β contributes to the repairing process of injured alveolar epithelium [486].

In the intensive care unit, acute lung injury is a well-recognized sequela of sepsis and systemic inflammatory response syndrome, caused by overwhelming host defense against infection, or by invasion of sterile tissues and fluids during traumatic injury or major surgery [81, 199]. Activation of the innate immune system is initiated by pattern recognition receptors (PRRS) that recognize specific structures of microbes or foreign bodies. Three families of PRRs are implicated in the response to septic insults or to sterile tissue injury: toll-like receptors (TLRs), nucleotide-oligomerization domain (NOD) leucine-rich repeat proteins, and retinoic acid-inducible gene I (RIG-I)-like helicases [81]. Using TLRs as an example, Gram-positive and Gram-negative bacteria, virus, and fungi have unique cell-wall molecules known as pathogen-associated molecular patterns (PAMPs), which bind to TLRs on the surface of immune cells, triggering their activation and further action directed at eradication of the pathogens (e.g., phagocytosis). Leukocytes are considered to be double-edged swords in this process because they are the major source of the cytokine storm (TNF-α and IL-1 production), and overproduction of oxidants and proteases, which not only destroy pathogens but also damage the structure and function of normal tissues and organs, especially the lungs.

Sepsis is also characterized by exacerbated coagulation, impaired anticoagulation, and decreased fibrin removal [81]. Many pro-inflammatory mediators produced during the early stages of sepsis are also pro-coagulatory. For example, thrombin, a well-known permeability-increasing factor, plays an essential role in blood clot formation. Another pro-inflammatory cytokine, interleukin-6, induces tissue factor upregulation thereby activating the extrinsic coagulatory cascade. In addition, tumor necrosis factor-alpha is known to suppress natural anticoagulants, such as activated protein C. These factors may act in concert shifting the hemostatic balance toward coagulation. Thrombosis occurs and microthrombi obstruct the microcirculation, impairing local tissue perfusion and exchange. Microthrombosis causes further damage to the endothelial barrier by promoting oxidative stress and apoptosis in endothelial cells. Fibrinolytic products released from the clots can also increase microvascular permeability by directly binding to the integrins expressed on the endothelial cell surface [170].

Recombinant human activated protein C, Drotrecogin-alpha (DrotAA), is the first anticoagulatory agent that was proven to be effective in the treatment of sepsis. As described in the previous chapter, APC is an endogenous anticoagulant with anti-inflammatory and antiapoptotic function [110, 399]. In addition to the well-described inhibitory effects on immune cells and inflammatory mediators, APC may exert its protection on endothelial barriers at the cellular level by competing with thrombin-induced cytoskeletal contraction and intercellular gap formation [212].

## ISCHEMIA–REPERFUSION INJURY

Ischemia–reperfusion (I/R) injury refers to tissue damage that occurs after blood supply returns to previously ischemic areas. I/R injury is typically observed under the clinical conditions of myocardial infarction, organ transplantation, and hemorrhage after resuscitation. I/R injury is often accompanied by oxidative stress due to an overproduction of reactive oxygen species (ROS), which damage endothelial barrier integrity and disrupt endothelial cell–cell junctions leading to microvascular hyperpermeability [153, 159].

ROS are multiple species generated by partial reduction of molecular oxygen during normal cell metabolism. ROS include superoxide ($O_2^\cdot$), hydroxyl ($OH^\cdot$), hydroperoxyl ($HO_2^\cdot$), peroxyl ($RO_2^\cdot$), alkoxyl ($RO^\cdot$) free radicals, and strong oxidizers: hydrogen peroxide ($H_2O_2$), hypochlorous acid ($HOCl$), hydroperoxide ($ROOH$), and lipid peroxides [55]. Under pathophysiological conditions, excess ROS are produced by mitochondrial respiration, NADPH oxidase, xanthine oxidase, cytochrome P450, cyclooxygenase (COX) and activated eNOS [64, 270]. Excess accumulation of ROS in response to increased ROS production, or decreased ROS metabolism by superoxide dismutases (SODs), causes cellular oxidative stress and tissue injury by impairing cell membranes, proteins, metabolism and signal transduction [55, 73]. ROS affects cell signaling by directly modifying cellular proteins, or by combining with excess NO to form the highly reactive species peroxynitrite ($ONOO^-$), and reactive byproducts: ($ONOOH$), nitrate or nitrite. Under hypercapnic conditions (as occurs in ischemia) $ONOOH$ can combine with $CO_2$ to form the reactive species $ONOOCO_2^-$ [173]. Thus, oxidative stress generates many reactive species in endothelial cells that can induce barrier dysfunction.

In pulmonary endothelium, excess ROS causes activation of the transient receptor potential melastatin (TRPM)-2 cation channel, and $Ca^{2+}$ influx, through activation of TRPM2 by PKC-$\alpha$ [186–188]. In brain microvascular endothelium, ROS caused hyperpermeability through PLC activation, and $Ca^{2+}$-dependent activation of ERK1/2 [138]. Oxidative stress also causes activation of RhoA [55], and calmodulin kinase II [63, 65], indicating that endothelial hyperpermeability involves actin cytoskeleton reorganization and MLCK-dependent actomyosin contractility.

ROS further contribute to the progress of I/R injury by upregulating other inflammatory mediators, such as leukotriene $B_4$, thromboxane $A_2$, and adhesion molecules that mediate leukocyte adhesion and chemotaxis [475]. Oxidative stress also increases VEGF production and VEGF receptor expression in endothelial cells [13]. In the case of myocardial infarction, although upregulated

VEGF signaling benefits collateral microvessel formation, VEGF can aggravate pulmonary edema and cerebral hemorrhagic transformation [5, 80]. In the gastrointestinal system following ischemia, mucosal mast cells are activated by free radicals and subsequently release histamine thus inducing microvascular permeability and leukocyte infiltration [160]. Another pathway leading to I/R injury is complement activation, which not only upregulates pro-inflammatory cytokines but also activates leukocytes and endothelial cells [333]. As a consequence, these cells further release more oxidants, vasoactive factors, and proteases (elastases and cathepsin G) that directly target the microvascular barrier, resulting in plasma leakage and tissue edema [67, 88].

ROS can be produced in activated leukocytes, endothelial cells as well as resident macrophages in tissues; however, activated leukocytes are the most important source of ROS and barrier-damaging proteases during I/R injury. Neutrophils are generally the first cells to arrive at the site of injury. Neutrophil interactions with endothelial adhesion molecules induce neutrophil activation [112]. Fully activated neutrophils undergo a respiratory burst, characterized by degranulation and release of numerous hyperpermeability-inducing agents, including enzymes (e.g., myeloperoxidase, elastase, cathepsin G, lysozyme, proteases, alkaline phosphatase, etc.), arachidonic acid, and ROS [55, 112]. Many of these compounds induce endothelial barrier dysfunction, including cathepsin G (cleaves VE–cadherin), and arachidonic acid (AA) metabolites: leukotrienes (LTs), thromboxanes (TXs), and prostaglandins (PGEs), known inducers of endothelial hyperpermeability. In general, AA is produced by liberation from the membrane in response to activation of PLC by activated G-proteins. AA is in turn converted to PGEs and TXs by cyclooxygenases, and to LTs by lipoxygenases [112].

During infection or inflammation, activated leukocytes extravasate from the blood into the surrounding tissue (diapedesis) [126]. Activated and firmly adhered leukocytes induce changes in endothelial cell morphology, which allow leukocyte transmigration [125, 126, 166, 192, 268]. In particular, leukocyte adhesion at adherens junctions displaces VE–cadherin. Other studies show that VE–cadherin is retracted during leukocyte transmigration forming a gap through which the leukocyte can traverse the cell–cell junction [288]. It is proposed that neutrophils secrete proteases that cleave VE–cadherin. However, the assumption that gap formation is required for leukocyte diapedesis is challenged by the observation that endothelial permeability is not increased during transmigration [234, 266]. Other investigators have suggested that adhesion molecules on the surface of leukocytes can temporarily bind to intercellular junction proteins on endothelial cells (i.e., PECAM-1) and disrupt cell–cell adhesions sequentially in a zipper-like fashion, without increasing junction permeability. A model of this kind is supported by electron micrographs showing neutrophils migrating between endothelial cells in the absence of albumin leakage [267], as well as fluorescence microscopy images showing leukocytes in very close contact with endothelial cell membranes prior to and during diapedesis [68, 69]. Further studies of neutrophil diapedesis during aseptic dermal injury demonstrate a clear separation between the time of endothelial transmigration and that of any measurable increase in endothelial permeability [234].

In summary, leukocyte transendothelial migration is often accompanied by endothelial hyperpermeability. Whether the two processes are merely associated events or have a cause–effect relationship remains a controversial issue. In general, the detrimental effects of leukocytes on microvascular endothelium during reperfusion are well recognized, and the mechanisms of leukocyte-induced barrier injury are largely attributed to their overall activation status (producing ROS and proteases) and adhesion to endothelium (increasing the local concentration of leukocyte-derived factors and adhesion molecule-triggered intracellular signaling).

# DIABETES MELLITUS

Diabetes mellitus represents a group of metabolic/inflammatory disorders associated with defects in insulin production or utilization. The major cause of morbidity and mortality in diabetic patients is end-organ diseases manifest as retinopathy, nephropathy, neuropathy, cardiomyopathy, peripheral vascular disease, and foot disease. Most of these problems stem from microcirculatory disturbances. Increased capillary permeability is a hallmark of functional angiopathy that occurs before the onset of structural pathology characterized by capillary basement membrane thickening, microaneurysm, and neovascularization [421, 447]. The pathological consequence of microvascular hyperpermeability is deposition of plasma proteins and lipids in the vascular wall, altering matrix metabolism and contributing to the morphogenesis of microangiopathic lesions and hypertension.

Hyperglycemia is considered the most important etiological factor in diabetic microvascular disease. Clinical studies have established a causal relationship between high levels of blood glucose and glycated hemoglobin with the prevalence of nephropathy, retinopathy, and coronary and cerebral microvascular diseases [60, 185]. In vivo analyses of microvascular transport demonstrate an increased transvascular flux of albumin and lipoproteins in patients with hyperglycemia, regardless of the sub-types of diabetes [108, 241, 453]. Leakage of blood components is detected in the heart, kidney, brain, retina, skin, and skeletal muscle of animals in which hyperglycemia is induced by chemicals or surgical ablation of pancreas beta-cell production of insulin [37, 164, 207, 298, 453, 500, 512]. In cultured endothelial cells, glucose and its metabolites induce concentration-dependent macromolecular flux [181, 191, 332]. Control experiments with L-glucose or manitol indicate that the hyperpermeability is unlikely due to osmotic damage of the endothelial barrier; rather, D-glucose may alter endothelial barrier integrity by causing a series of biochemical and metabolic reactions in the endothelium.

As previously discussed, plasma transport across microvascular endothelium is controlled by physical forces and the barrier property of the endothelium. Under diabetic conditions, both altered microcirculatory hemodynamics and endothelial hyperpermeability contribute to imbalanced fluid homeostasis. Intravascular hydrostatic pressure is the major force driving the outward flux of blood fluid and macromolecules. Vasodilation upstream of the microvasculature can result in elevated perfusion pressure in the downstream exchange microvascular bed, facilitating plasma extravasation. In diabetic patients and animals, chronic vasodilation accompanied by a high resting blood flow and reduced autoregulatory capacity is commonly seen in the peripheral microcirculation [31, 52, 84, 85]. In fact, increased renal perfusion is considered the major factor responsible for excessive glomerular filtration and microalbuminuria in diabetic nephropathic patients [421]. In the retina, dilated and leaky microvessels represent the morphological characteristics of diabetic retinopathy [9, 66, 100]. Additionally, increased basal coronary flow and reduced vasodilator reserve are seen in both types of diabetes before the occurrence of clinically significant coronary artery disease [298]. Thus, the hyperperfusion state accounts, at least in part, for the increases in albumin transport observed in vivo under hyperglycaemic conditions.

Another important mechanism underlying diabetic microvascular disease is increased endothelial permeability. Endothelial fenestration (large intercellular gaps) is considered an ultrastructural hallmark of microangiopathy [84, 164, 298, 421]. Impaired tight junction structures are seen in diabetic retina and cerebral microvessels [23, 76]. The expression of adherens junctions is also reduced in the retinal microvasculature of diabetic patients [102]. While there are many reports on the correlation between endothelial cytoskeleton contractile morphology and paracellular leakage [388, 506], an electron microscopic study in the rat coronary microvasculature reveals more luminal vesicles containing gold-labeled albumin after STZ treatment [500], suggesting enhanced transcellular transport. Likewise, a comparative analysis of magnetic resonance imaging with ultrastructural immunolocalization indicates that the blood-brain barrier breakdown in diabetic rabbits is principally mediated by the transendothelial transport of proteins in endocytotic vesicle-like structures [467]. Taken together, these results lead to the agreement that both paracellular and transcellular pathways contribute to diabetic microvascular hyperpermeability.

The pathogenic factors of diabetic microvascular complications include advanced glycation, oxidative stress, and PKC abnormalities [60]. Irreversible advanced glycation end products (AGEs) are produced by autooxidation and nonenzymatic glycation of proteins or lipids [162, 400]. The process mainly involves the Maillard reaction, where aldehydes or ketones of reducing sugar bind to free amino groups of proteins forming a labile Schiff base and producing Amadori products. Amadori can be degraded into a variety of highly reactive carbonyl compounds capable of cross-linking molecules, leading to carbonyl stress. AGEs are found in the blood and vasculature of diabetic patients, particularly in those with poorly controlled glycemia [162, 391]. Evidence is emerging that AGEs induce microvessel hyperpermeability via three mechanisms. First, AGEs activate multiple inflammatory signaling cascades via binding to their receptors, RAGEs, which are highly expressed in vascular endothelium [400]. In vivo studies suggest that increased solute flux in diabetic animal models can be prevented by blockage of RAGE-AGE interactions with either neutralizing anti-bodies or soluble RAGEs [54, 479]. Second, glycation of intracellular molecules involved in the regulation of gene transcription, such as NF-κB and MAP kinases, leads to abnormal production of pro-inflammatory cytokines and growth factors [60], which in turn can cause hyperpermeability. The third possibility is that glycosylated plasma proteins alter the chemical or physical structure of the negatively charged fibrous meshwork on the endothelial surface, the glycocalyx, which is be-lieved to play a major role in maintaining the size-selective sieve property of exchange microvessels [94, 466].

In addition to carbonyl stress, glucose metabolites and free fatty acids may alter endothelial barrier function by causing oxidative stress via stimulation of mitochondrial respiration, alteration of redox states, and nitric oxide activity uncoupling in the microvascular endothelium [432, 441]. In particular, excessive sorbitol production via the polyol (aldose reductase) pathway results in a consumption of NADPH and a decrease in the level of reduced glutathione [60]. Free fatty acids can stimulate oxidative phosphorylation in the mitochondria producing superoxide [403]. Also, AGEs activate NADPH oxidase, the major source of superoxide in endothelial cells [441]. In ani-mal models of diabetes, treatment with pharmacological doses of antioxidants, such as vitamins C and E, prevents vascular and neurological dysfunction [185]. Unfortunately, clinical trials with these antioxidants have not demonstrated prevention of renal or cardiovascular dysfunction in diabetic patients [185]. The discrepant outcomes between human and animal studies underscore the need for further comparative analyses at the molecular level.

Animal experiments suggest a causal role for PKC activation in diabetic microangiopathy, as selective inhibition of PKC normalizes endothelial function and prevents microvascular leakage in the diabetic retina and kidney [9, 243]. Yuan's lab has previously reported that PKCβ activation in porcine hearts occurs during the early stages of diabetes concomitantly with coronary venular hyperpermeability that can be reversed by selective inhibition of PKCβ [172, 512]. The therapeutic benefit of PKC blockade is exemplified by recent clinical trials reporting that the PKCβ inhibitor ruboxistaurin (LY333531) delays macular edema progression in retinopathic patients [217]. The drug also normalizes glomerular filtration and reduces microalbuminuria in diabetic nephropathic patients [449]. Furthermore, PKC has been linked to insulin resistance based on its capability to modulate insulin receptor phosphorylation and PI3K-Akt signaling [128, 371].

Despite the well-studied downstream effects of PKC, little is known about the causes of PKC activation in diabetic conditions. A recent hypothesis [100, 210] points to the de novo synthesis of DAG as a potential mechanism. Indeed, the production of DAG is elevated in microvessels and endothelial cells exposed to high glucose [488]. In a porcine model of diabetes, an increased level of DAG was detected in the blood and cardiovascular tissues, correlating with elevated PKC activity and coronary venular hyperpermeability [512]. However, while the de novo synthesis of DAG can lead to increased PKC activity, it may not be sufficient to explain the well-documented increase in PKC abundance in diabetic tissues [221, 278]. In this regard, a recent gene analysis in diabetic pig hearts and vessels showed overexpression of PKCβ at the mRNA and protein levels [172], suggesting that PKC upregulation occurs during diabetes. In line with this notion, evidence exists supporting a role for glucose in promoting gene expression and protein synthesis through activation of MAP kinases and other transcription factors [430, 446].

Diabetic retinopathy represents a typical form of microvascular complication in diabetes mellitus. Its pathophysiology is characterized by retinal microvascular hyperpermeability, aneurysm, and hemorrhage [9, 66, 100, 243]. Blood components leak into the tissue, forming exudate deposits that block the macula and impair vision. The disease advances into the proliferative stage where new blood vessels are formed and fibrovascular tissues grow into the vitreous cavity, leading to retinal detachment and blindness. VEGF is thought to be the primary mediator of both the nonproliferative and proliferative phases of the disease.

VEGF is a family of growth factors capable of inducing new blood vessel formation (angiogenesis) and microvascular hyperpermeability. VEGF consists of VEGF A-E, and placental growth

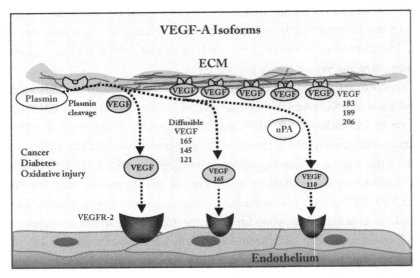

**FIGURE 31:** VEGF-A isoforms in the endothelial extracellular matrix (ECM). The ECM serves as a reservoir to sequester multiple VEGF-A isoforms of varying size. During diabetes, cancer or oxidative stress, VEGF-A isoforms are released from the ECM following cleavage by proteases (e.g., plasmin, MMPs). Smaller isoforms (VEGF-121, -145 or -165) are more readily diffusible than larger isoforms; Larger isoforms (-183, -189 or -206) must be cleaved into smaller (-110) active fragments by urokinase type plasminogen activator (uPa), to freely dissociate from the ECM. VEGF isoforms bind to VEGF receptors (VEGFR) on the surface of endothelium: binding to VEGFR-2 (KDR) causes hyperpermeability.

factor [137]. Of these, VEGF-A is centrally important for angiogenic and permeability responses (Figure 31). VEGF-A exists as multiple isoforms (VEGF-121, -145, -165, -183, -189, and -206) due to splice variation, of which VEGF-165 is the most abundant in human tissue. VEGF-A isoforms bind to heparin sulfate proteoglycans, present on the surface of endothelium and throughout the ECM. VEGF isoforms released from the ECM are freely diffusible, with the smaller isoforms being most able to diffuse. The larger isoforms [189 and 206] are less diffusible and are more often distributed by secretion from cells and then are fully retained within the ECM. VEGF is released from the ECM in response to cleavage by the serine protease, plasmin. Plasmin cleaves VEGF-A isoforms into a 110 amino acid N-terminal fragment that can no longer bind to heparin sulfate, and that is capable of binding to VEGFR-2 on the surface of endothelial cells. The longer VEGF isoforms [183, 189, and 206] contain an additional site for cleavage by urokinase type plasminogen activator (uPA), which produces a similarly active fragment of VEGF.

VEGF causes microvascular hyperpermeability by directly targeting the endothelium via a receptor (KDR)-dependent mechanism [492, 497]. The activation of this receptor tyrosine kinase triggers multiple intracellular signals, including PLC, PKC, eNOS, MAPKs, and the PI3K-dependent Akt/PKB pathway [24, 32, 33, 491, 492, 497]. A downstream effect of these signaling reactions is focal adhesion-supported cytoskeletal contraction, coupled with the opening of intercellular junctions [436, 493]. Other studies have shown that VEGF down-regulates the expression of tight junctions and induces junctional molecule sequestration [66]. In addition, VEGF is known to promote plasma protein transcytosis in endothelial cells; this effect may involve eNOS activation and intracellular translocation [394, 508].

Experimental evidence suggests that both VEGF production and VEGF receptor expression are increased in the diabetic retina, and that inhibition of VEGF signaling prevents blood–retina barrier breakdown and suppresses neovascularization [66]. The large volume of supporting data stimulates the enthusiasm for developing anti-VEGF therapies to treat diabetic retinopathy. In patients with diabetic macular edema, the therapeutic effect of anti-VEGF agents has been tested using fluorescein angiograph and optical coherence tomography [382]. Subsequently, a number of VEGF receptor antagonists have been developed and are currently under preclinical or clinical trials. They act either as pan-inhibitors of VEGF receptors (e.g., pazopanib), or as specific inhibitors of KDR (e.g., ZM323881). In animal studies, ZM323881 effectively blocks the acute vascular leak response caused by VEGF [178, 485].

In addition to the strategies specifically targeting VEGF/KDR, the barrier-protective effect of opposing VEGF-induced damage could be derived from other approaches including administration of angiopoietin-1 (Ang-1) or inhibition of PLC, PKC, PKG, intracellular $Ca^{2+}$, MEK or Src [485, 492, 497, 510]. Although both Ang-1 and VEGF are proangiogenic factors, Ang-1 promotes vessel maturation and inhibits VEGF-induced endothelial hyperpermeability. A recent study suggests that the protective effect of Ang-1 is mediated through inhibition of Src kinase activation

[458]. In support of this, it has been demonstrated that VEGF-induced vascular barrier dysfunction is specifically blocked in Src-deficient-mice while sparing normal angiogenesis [122]. Further evidence shows that topical application of a novel KDR/Src kinase inhibitor suppresses VEGF-mediated retinal vascular hyperpermeability in animal models [151].

## TUMOR ANGIOGENESIS AND METASTATIC MIGRATION

Tumor growth requires excessive supply of oxygen and nutrients and active removal of metabolic wastes. This extremely high level of blood–tissue exchange is supported by a dense microvasculature developed through pathological angiogenesis. Angiogenesis occurs in the early stages of carcinogenesis in response to VEGF and bFGF produced by the tumor or stromal cells. Unlike other microvessels, tumor-induced neovasculature possesses a more fenestrated endothelium, and most of tumor microvessels are thin-walled with leaky endothelium, partially due to underdeveloped endothelial cell–cell junctions and discontinuous smooth muscles and pericytes [66, 182]. Thus, they are more permeable and possibly more conducive to tumor cell intravasation.

The hyperpermeability property of the endothelium plays a crucial role in the initial development and continued growth of tumors, as it facilitates plasma accumulation in the matrix to support new vessel formation. This process is characterized by a highly orchestrated cellular response requiring interactions of multiple growth factors, adhesion molecules, and matrix proteins [316]. Typically, VEGF is considered a key signal for endothelial barrier breakdown that allows endothelial cell migration and matrix-supported capillary growth. In breast cancer and other types of cancer, an excessive amount of VEGF is detected [353] and its expression level correlates with the malignant degree of the tumor [21], whereas inhibition of VEGF signaling suppresses tumor angiogenesis and malignant progression [120, 259]. Previously, it has been demonstrated that VEGF increases endothelial permeability and promotes angiogenesis via complex endothelial cell–cell and cell–matrix interactions [507].

Based on the importance of VEGF in angiogenesis-associated diseases (tumorigenesis and diabetic retinopathy), tremendous research effort has been devoted to developing novel strategies that inhibit the VEGF pathway. Current products mainly include anti-VEGF antibodies, VEGF traps (genetically engineered soluble VEGF receptors), and VEGF receptor blockades. Animal experiments demonstrate that VEGF neutralizing antibodies can suppress microvascular permeability and angiogenesis in tumors thereby reducing tumor size [259]. Bevacizumab, the first FDA-approved anti-VEGF agent, not only inhibits tumor growth in a dose-dependent fashion [120] but also produces an antimetastatic effect [120, 155]. This drug has been used along with chemotherapy in patients with lung, renal, or colorectal cancer.

Metastasis of tumor cells to distant sites is a complex, multi-step consequence of tumor progression [232]. The metastatic cascade can be described as a series of events: detaching from the primary tumor mass; penetrating the basement membrane and invading the surrounding stroma;

intravasation into the circulation; transportation in the vascular system while evading from host immune surveillance; attachment to the vascular endothelium of target organs and extravasation; and finally establishment of new growth (Figure 32). Most of these steps involve tumor cell movement (across vascular endothelium and surrounding matrix) assisted by activities of proteases, such as MMPs and ADAMs, which not only degrade the basement membrane and stromal extracellular matrix, but also cleave and activate cytokines/growth factors promoting tumor cell invasion and migration [283, 326].

Endothelial barrier dysfunction plays an important role in metastatic cancer development. In addition to providing the structural basis for angiogenesis, endothelial cell–cell junctions serve as

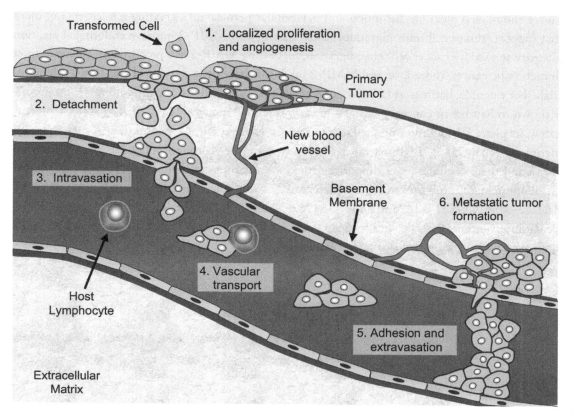

**FIGURE 32:** The multiple steps of tumor metastasis. The sequential steps of tumor metastasis include (1) proliferation and transformation of cells in the primary tumor, (2) detachment and migration of metastatic tumor cells, (3) penetration of the endothelial basement membrane and intravasation into the circulation, (4) transport and evasion of immune responses within the blood, (5) attachment to the endothelium and extravasation into the tissues of target organs, and (6) establishing new tumor growth.

the major route for tumor cell transmigration (TEM) across the microvascular wall during metastasis via the blood circulation. Currently, many theories consider tumor cell TEM a process similar to leukocyte extravasation during inflammation; the sequence of events includes circulating cell adhesion to the endothelium, endothelial cell retraction and junction opening, followed by cell transmigration. Therefore, the barrier function of the vascular endothelium is an important determinant of transmigration and extravasation of cells from the circulation. In co-culture studies where breast cancer cells are grown on top of human lung microvascular endothelial cell monolayers [222], cancer cell transmigration occurs through intercellular spaces where cell–cell junctions (VE–cadherin) are loosened or sequestrated. The transmigration is coupled with decreased transendothelial electrical resistance, indicating endothelial barrier dysfunction. Further molecular analyses demonstrate that metastatic cell adhesion to lung endothelial cells induces MMP-2 activation, which in turn causes endothelial junction disruption and paracellular permeability, creating a microenvironment that supports tumor cell transmigration. Antibodies against MMP-2 improve endothelial junction integrity and reduce cancer cell transmigration, suggesting MMP-2 as a potential therapeutic target in metastatic cancer. Indeed, several MMP-2 inhibitors have been developed and tested in clinical trials. For example, batimastat (BB-94), one of the early generations of peptidomimetic inhibitors, is shown to halt breast cancer metastasis in various animal models [121, 214, 426, 498]. Unfortunately, its phase III trail is terminated due to severe adverse reactions in local tissues. Subsequently, marimastat (BB-2516), a batimastat derivative that can be delivered orally, has been developed [193], and there is clinical data supporting the use of marimastat in advanced cancer [56]. Other metalloproteinase inhibitors that have been tested as anticancer therapies include BMS-275291, BAY 12-9566, CGS27023A, and FYK-1388 [325, 346, 484]. Further development of drugs or molecular interventions that specifically target the endothelial junction barrier may serve as a potential alternative means for blockage of tumor metastasis.

.   .   .   .

# Conclusion

Endothelial barriers are critical for maintaining fluid/electrolyte homeostasis and physiological organ function. The microvascular endothelium experiences diverse stimulation in the form of physical forces, chemical factors, and circulating cells, and its permeability is tightly regulated by moment-to-moment interactions between endothelial cell–cell and cell–matrix adhesion molecules. As the initiator and consequence of microvascular inflammation, endothelial barrier dysfunction is a critical problem underlying the development of many diseases or complications of injury. Over the past few decades, tremendous progress has been made in animal research and cell/molecular biology experiments have led to an in-depth understanding of the pathophysiological mechanisms underlying endothelial hyperpermeability. However, translation of bench research data into clinical practice remains challenging. Of particular concern is the tremendous heterogeneity among species (animals vs. humans), tissues/organs (e.g., lungs vs. gut), and vessel segments (macrovessels vs. microvessels). Although some permeability-altering agents or mediators share similar intracellular signaling pathways, their temporal effects and patterns of resolution are rather different. This heterogeneity may be attributed to differential regulation of receptor binding, trafficking, internalization or desensitization. It is also possible that different levels of counter-regulatory signaling or feedback controls are activated concurrently with the initial hyperpermeability signaling. Further, despite the efficacy of antagonizing individual hyperpermeability-inducing mediators in certain pathological states, only moderate effectiveness is demonstrated under more complicated conditions where multiple mediators are produced and multiple organ systems are involved. Therefore, targeting their common terminal effectors may represent a more promising strategy against barrier injury. Within this context, site-specific drug delivery needs to be developed. A good strategy should improve the barrier function in affected vascular beds while sparing other tissues from unwanted effects. A similar issue applies to treating cancers, where microvascular hyperpermeability may favor the delivery of therapeutic drugs or diagnostic tools. Research efforts shall be devoted to further understanding the endothelial barrier structure and function for improved treatment of vascular inflammation.

# References

[1]     Abbott NJ. Role of intracellular calcium in regulation of brain endothelial permeability. In: *Introduction to the blood–brain barrier : methodology, biology and pathology*, edited by Pardridge WM. Cambridge, New York: Cambridge University Press, 1998: pp. 345–53.

[2]     Abbott NJ, Patabendige AA, Dolman DE, Yusof SR, and Begley DJ. Structure and function of the blood–brain barrier. *Neurobiol Dis* 37: pp. 13–25, 2010.

[3]     Abedi H and Zachary I. Vascular endothelial growth factor stimulates tyrosine phosphorylation and recruitment to new focal adhesions of focal adhesion kinase and paxillin in endothelial cells. *J Biol Chem* 272: pp. 15442–451, 1997.

[4]     Abram CL and Lowell CA. The ins and outs of leukocyte integrin signaling. *Annu Rev Immunol* 27: pp. 339–62, 2009.

[5]     Abumiya T, Yokota C, Kuge Y, and Minematsu K. Aggravation of hemorrhagic transformation by early intraarterial infusion of low-dose vascular endothelial growth factor after transient focal cerebral ischemia in rats. *Brain Res* 1049: pp. 95–103, 2005.

[6]     Adams RA, Schachtrup C, Davalos D, Tsigelny I, and Akassoglou K. Fibrinogen signal transduction as a mediator and therapeutic target in inflammation: lessons from multiple sclerosis. *Curr Med Chem* 14: pp. 2925–936, 2007.

[7]     Adamson RH, Lenz JF, Zhang X, Adamson GN, Weinbaum S, and Curry FE. Oncotic pressures opposing filtration across non-fenestrated rat microvessels. *J Physiol* 557: pp. 889–907, 2004.

[8]     Ahrns KS. Trends in burn resuscitation: shifting the focus from fluids to adequate endpoint monitoring, edema control, and adjuvant therapies. *Crit Care Nurs Clin North Am* 16: pp. 75–98, 2004.

[9]     Aiello LP. The potential role of PKC beta in diabetic retinopathy and macular edema. *Surv Ophthalmol* 47 Suppl 2: pp. S263–69, 2002.

[10]    Aitken A. Protein consensus sequence motifs. *Methods Mol Biol* 211: pp. 465–85, 2003.

[11]    Aitken A. Protein consensus sequence motifs. *Mol Biotechnol* 12: pp. 241–53, 1999.

[12]    Akakura N, Hoogland C, Takada YK, Saegusa J, Ye X, Liu FT, Cheung AT, and Takada Y. The COOH-terminal globular domain of fibrinogen gamma chain suppresses angiogenesis and tumor growth. *Cancer Res* 66: pp. 9691–7, 2006.

[13]    Aktan A and Yalcin AS. Ischemia–reperfusion injury, reactive oxygen metabolites, and the surgeon. *Turkish Journal of Medical Sciences* 28: pp. 1–5, 1998.

[14]    Albelda SM, Daise M, Levine EM, and Buck CA. Identification and characterization of cell–substratum adhesion receptors on cultured human endothelial cells. *J Clin Invest* 83: pp. 1992–2002, 1989.

[15]    Alcaide P, Newton G, Auerbach S, Sehrawat S, Mayadas TN, Golan DE, Yacono P, Vincent P, Kowalczyk A, and Luscinskas FW. p120-Catenin regulates leukocyte transmigration through an effect on VE–cadherin phosphorylation. *Blood* 112: pp. 2770–9, 2008.

[16]    Alexander JS, Jackson SA, Chaney E, Kevil CG, and Haselton FR. The role of cadherin endocytosis in endothelial barrier regulation: involvement of protein kinase C and actin-cadherin interactions. *Inflammation* 22: pp. 419–33, 1998.

[17]    Alexander JS, Patton WF, Christman BW, Cuiper LL, and Haselton FR. Platelet-derived lysophosphatidic acid decreases endothelial permeability in vitro. *Am J Physiol* 274: pp. H115–122, 1998.

[18]    Allard MF, Doerschuk CM, Brumwell ML, Belzberg A, and Hogg JC. Oleic acid-induced lung injury in rabbits: effect of fibrinogen depletion with Arvin. *J Appl Physiol* 64: pp. 920–8, 1988.

[19]    Allen TH, Ochoa M, Jr., Roth RF, and Gregersen MI. Spectral absorption of T-1824 in plasma of various species and recovery of the dye by extraction. *Am J Physiol* 175: pp. 243–6, 1953.

[20]    Allport JR, Ding H, Collins T, Gerritsen ME, and Luscinskas FW. Endothelial-dependent mechanisms regulate leukocyte transmigration: a process involving the proteasome and disruption of the vascular endothelial–cadherin complex at endothelial cell-to-cell junctions. *J Exp Med* 186: pp. 517–27, 1997.

[21]    Anan K, Morisaki T, Katano M, Ikubo A, Kitsuki H, Uchiyama A, Kuroki S, Tanaka M, and Torisu M. Vascular endothelial growth factor and platelet-derived growth factor are potential angiogenic and metastatic factors in human breast cancer. *Surgery* 119: pp. 333–9, 1996.

[22]    Andersen B and Ussing HH. Solvent drag on non-electrolytes during osmotic flow through isolated toad skin and its response to antidiuretic hormone. *Acta Physiol Scand* 39: pp. 228–39, 1957.

[23]    Antonetti DA, Lieth E, Barber AJ, and Gardner TW. Molecular mechanisms of vascular permeability in diabetic retinopathy. *Semin Ophthalmol* 14: pp. 240–8, 1999.

[24]    Aramoto H, Breslin JW, Pappas PJ, Hobson RW, 2nd, and Duran WN. Vascular endothelial growth factor stimulates differential signaling pathways in in vivo microcirculation. *Am J Physiol Heart Circ Physiol* 287: pp. H1590–8, 2004.

[25]    Armenante PM, Kim D, and Duran WN. Experimental determination of the linear cor-relation between in vivo TV fluorescence intensity and vascular and tissue FITC-DX con-centrations. *Microvasc Res* 42: pp. 198–208, 1991.

[26]    Arturson G. Forty years in burns research—the postburn inflammatory response. *Burns* 26: pp. 599–604, 2000.

[27]    Arturson G. Pathophysiology of the burn wound and pharmacological treatment. The Rudi Hermans Lecture, 1995. *Burns* 22: pp. 255–74, 1996.

[28]    Augustin HG, Koh GY, Thurston G, and Alitalo K. Control of vascular morphogenesis and homeostasis through the angiopoietin-Tie system. *Nat Rev Mol Cell Biol* 10: pp. 165–177, 2009.

[29]    Avraham HK, Lee TH, Koh Y, Kim TA, Jiang S, Sussman M, Samarel AM, and Avraham S. Vascular endothelial growth factor regulates focal adhesion assembly in human brain microvascular endothelial cells through activation of the focal adhesion kinase and related adhesion focal tyrosine kinase. *J Biol Chem* 278: pp. 36661–8, 2003.

[30]    Barnes DW, Loutit JF, and Reeve EB. Observations on the estimate of the circulating red blood cell volume in man given by T 1824 and the haematocrit, with special reference to uncorrected dye loss from the circulation. *Clin Sci (Lond)* 7: pp. 155–173, 1948.

[31]    Baron AD. The coupling of glucose metabolism and perfusion in human skeletal muscle. The potential role of endothelium-derived nitric oxide. *Diabetes* 45 Suppl 1: pp. S105–9, 1996.

[32]    Bates DO. Vascular endothelial growth factors and vascular permeability. *Cardiovasc Res*, 2010.

[33]    Bates DO and Harper SJ. Regulation of vascular permeability by vascular endothelial growth factors. *Vascul Pharmacol* 39: pp. 225–37, 2002.

[34]    Bates DO, Levick JR, and Mortimer PS. Starling pressures in the human arm and their alteration in postmastectomy oedema. *J Physiol* 477 ( Pt 2): pp. 355–63, 1994.

[35]    Bauer PM, Yu J, Chen Y, Hickey R, Bernatchez PN, Looft-Wilson R, Huang Y, Giordano F, Stan RV, and Sessa WC. Endothelial-specific expression of caveolin-1 impairs microvas-cular permeability and angiogenesis. *Proc Natl Acad Sci U S A* 102: pp. 204–9, 2005.

[36]    Bauer TT, Monton C, Torres A, Cabello H, Fillela X, Maldonado A, Nicolas JM, and Zavala E. Comparison of systemic cytokine levels in patients with acute respiratory distress syndrome, severe pneumonia, and controls. *Thorax* 55: pp. 46–52, 2000.

[37]    Beals CC, Bullock J, Jauregui ER, and Duran WN. Microvascular clearance of macromol-ecules in skeletal muscle of spontaneously diabetic rats. *Microvasc Res* 45: pp. 11–19, 1993.

[38]    Becker BF, Chappell D, Bruegger D, Annecke T, and Jacob M. Therapeutic strategies tar-geting the endothelial glycocalyx: acute deficits, but great potential. *Cardiovasc Res*, 2010.

[39]    Bennett JS. Platelet-fibrinogen interactions. *Ann N Y Acad Sci* 936: pp. 340–54, 2001.

[40]   Berne RM, Koeppen BM, and Stanton BA. *Berne & Levy physiology*. Philadelphia: Mosby/Elsevier, 2008: pp. xii, 834.

[41]   Berridge MJ. Inositol trisphosphate and calcium signalling. *Nature* 361: pp. 315–25, 1993.

[42]   Bhatia M and Moochhala S. Role of inflammatory mediators in the pathophysiology of acute respiratory distress syndrome. *J Pathol* 202: pp. 145–56, 2004.

[43]   Bianchi E, Bender JR, Blasi F, and Pardi R. Through and beyond the wall: late steps in leukocyte transendothelial migration. *Immunol Today* 18: pp. 586–91, 1997.

[44]   Birukov KG. Small GTPases in mechanosensitive regulation of endothelial barrier. *Microvasc Res* 77: pp. 46–52, 2009.

[45]   Birukov KG, Csortos C, Marzilli L, Dudek S, Ma SF, Bresnick AR, Verin AD, Cotter RJ, and Garcia JG. Differential regulation of alternatively spliced endothelial cell myosin light chain kinase isoforms by p60(Src). *J Biol Chem* 276: pp. 8567–73, 2001.

[46]   Birukova AA, Birukov KG, Smurova K, Adyshev D, Kaibuchi K, Alieva I, Garcia JG, and Verin AD. Novel role of microtubules in thrombin-induced endothelial barrier dysfunction. *FASEB J* 18: pp. 1879–90, 2004.

[47]   Birukova AA, Liu F, Garcia JG, and Verin AD. Protein kinase A attenuates endothelial cell barrier dysfunction induced by microtubule disassembly. *Am J Physiol Lung Cell Mol Physiol* 287: pp. L86–93, 2004.

[48]   Birukova AA, Smurova K, Birukov KG, Usatyuk P, Liu F, Kaibuchi K, Ricks-Cord A, Natarajan V, Alieva I, Garcia JG, and Verin AD. Microtubule disassembly induces cytoskeletal remodeling and lung vascular barrier dysfunction: role of Rho-dependent mechanisms. *J Cell Physiol* 201: pp. 55–70, 2004.

[49]   Blackwell TS and Christman JW. The role of nuclear factor-kappa B in cytokine gene regulation. *Am J Respir Cell Mol Biol* 17: pp. 3–9, 1997.

[50]   Blobel CP. ADAMs: key components in EGFR signalling and development. *Nat Rev Mol Cell Biol* 6: pp. 32–43, 2005.

[51]   Bogatcheva NV, Garcia JG, and Verin AD. Molecular mechanisms of thrombin-induced endothelial cell permeability. *Biochemistry (Mosc)* 67: pp. 75–84, 2002.

[52]   Bohlen HG. Mechanisms for early microvascular injury in obesity and type II diabetes. *Curr Hypertens Rep* 6: pp. 60–5, 2004.

[53]   Boire A, Covic L, Agarwal A, Jacques S, Sherifi S, and Kuliopulos A. PAR1 is a matrix metalloprotease-1 receptor that promotes invasion and tumorigenesis of breast cancer cells. *Cell* 120: pp. 303–13, 2005.

[54]   Bonnardel-Phu E, Wautier JL, Schmidt AM, Avila C, and Vicaut E. Acute modulation of albumin microvascular leakage by advanced glycation end products in microcirculation of diabetic rats in vivo. *Diabetes* 48: pp. 2052–8, 1999.

[55] Boueiz A and Hassoun PM. Regulation of endothelial barrier function by reactive oxygen and nitrogen species. *Microvasc Res* 77: pp. 26–34, 2009.

[56] Bramhall SR, Hallissey MT, Whiting J, Scholefield J, Tierney G, Stuart RC, Hawkins RE, McCulloch P, Maughan T, Brown PD, Baillet M, and Fielding JW. Marimastat as maintenance therapy for patients with advanced gastric cancer: a randomised trial. *Br J Cancer* 86: pp. 1864–70, 2002.

[57] Breil I, Koch T, Belz M, Van Ackern K, and Neuhof H. Effects of bradykinin, histamine and serotonin on pulmonary vascular resistance and permeability. *Acta Physiol Scand* 159: pp. 189–98, 1997.

[58] Brouhard BH, Carvajal HF, and Linares HA. Burn edema and protein leakage in the rat. I. Relationship to time of injury. *Microvasc Res* 15: pp. 221–8, 1978.

[59] Brown MT and Cooper JA. Regulation, substrates and functions of src. *Biochim Biophys Acta* 1287: pp. 121–49, 1996.

[60] Brownlee M. The pathobiology of diabetic complications: a unifying mechanism. *Diabetes* 54: pp. 1615–25, 2005.

[61] Brugge JS and Erikson RL. Identification of a transformation-specific antigen induced by an avian sarcoma virus. *Nature* 269: pp. 346–8, 1977.

[62] Butcher EC. Leukocyte-endothelial cell recognition: three (or more) steps to specificity and diversity. *Cell* 67: pp. 1033–6, 1991.

[63] Cai H, Davis ME, Drummond GR, and Harrison DG. Induction of endothelial NO synthase by hydrogen peroxide via a Ca(2+)/calmodulin-dependent protein kinase II/janus kinase 2-dependent pathway. *Arterioscler Thromb Vasc Biol* 21: pp. 1571–6, 2001.

[64] Cai H and Harrison DG. Endothelial dysfunction in cardiovascular diseases: the role of oxidant stress. *Circ Res* 87: pp. 840–4, 2000.

[65] Cai H, Liu D, and Garcia JG. CaM kinase II-dependent pathophysiological signalling in endothelial cells. *Cardiovasc Res* 77: pp. 30–4, 2008.

[66] Caldwell RB, Bartoli M, Behzadian MA, El-Remessy AE, Al-Shabrawey M, Platt DH, Liou GI, and Caldwell RW. Vascular endothelial growth factor and diabetic retinopathy: role of oxidative stress. *Curr Drug Targets* 6: pp. 511–24, 2005.

[67] Carden DL and Granger DN. Pathophysiology of ischaemia-reperfusion injury. *J Pathol* 190: pp. 255–66, 2000.

[68] Carman CV, Sage PT, Sciuto TE, de la Fuente MA, Geha RS, Ochs HD, Dvorak HF, Dvorak AM, and Springer TA. Transcellular diapedesis is initiated by invasive podosomes. *Immunity* 26: pp. 784–97, 2007.

[69] Carman CV and Springer TA. A transmigratory cup in leukocyte diapedesis both through individual vascular endothelial cells and between them. *J Cell Biol* 167: pp. 377–88, 2004.

[70]    Carvajal HF, Linares HA, and Brouhard BH. Relationship of burn size to vascular permeability changes in rats. *Surg Gynecol Obstet* 149: pp. 193–202, 1979.

[71]    Cavanagh SP, Gough MJ, and Homer-Vanniasinkam S. The role of the neutrophil in ischaemia-reperfusion injury: potential therapeutic interventions. *Cardiovasc Surg* 6: pp. 112–8, 1998.

[72]    Cesarman-Maus G and Hajjar KA. Molecular mechanisms of fibrinolysis. *Br J Haematol* 129: pp. 307–21, 2005.

[73]    Chabot F, Mitchell JA, Gutteridge JM, and Evans TW. Reactive oxygen species in acute lung injury. *Eur Respir J* 11: pp. 745–57, 1998.

[74]    Chan JY, Takeda M, Briggs LE, Graham ML, Lu JT, Horikoshi N, Weinberg EO, Aoki H, Sato N, Chien KR, and Kasahara H. Identification of cardiac-specific myosin light chain kinase. *Circ Res* 102: pp. 571–80, 2008.

[75]    Charrier-Hisamuddin L, Laboisse CL, and Merlin D. ADAM-15: a metalloprotease that mediates inflammation. *FASEB J* 22: pp. 641–53, 2008.

[76]    Chehade JM, Haas MJ, and Mooradian AD. Diabetes-related changes in rat cerebral occludin and zonula occludens-1 (ZO-1) expression. *Neurochem Res* 27: pp. 249–52, 2002.

[77]    Cheng YF, Clyman RI, Enenstein J, Waleh N, Pytela R, and Kramer RH. The integrin complex alpha v beta 3 participates in the adhesion of microvascular endothelial cells to fibronectin. *Exp Cell Res* 194: pp. 69–77, 1991.

[78]    Chidlow JH, Jr. and Sessa WC. Caveolae, caveolins, and cavins: complex control of cellular signalling and inflammation. *Cardiovasc Res* 86: pp. 219–25, 2010.

[79]    Chien S, Li S, Shiu YT, and Li YS. Molecular basis of mechanical modulation of endothelial cell migration. *Front Biosci* 10: pp. 1985–2000, 2005.

[80]    Chua CC, Hamdy RC, and Chua BH. Upregulation of vascular endothelial growth factor by H2O2 in rat heart endothelial cells. *Free Radic Biol Med* 25: pp. 891–7, 1998.

[81]    Cinel I and Dellinger RP. Advances in pathogenesis and management of sepsis. *Curr Opin Infect Dis* 20: pp. 345–52, 2007.

[82]    Cioffi DL, Moore TM, Schaack J, Creighton JR, Cooper DM, and Stevens T. Dominant regulation of interendothelial cell gap formation by calcium-inhibited type 6 adenylyl cyclase. *J Cell Biol* 157: pp. 1267–78, 2002.

[83]    Cioffi WG. What's new in burns and metabolism. *J Am Coll Surg* 192: pp. 241–54, 2001.

[84]    Clark MG, Barrett EJ, Wallis MG, Vincent MA, and Rattigan S. The microvasculature in insulin resistance and type 2 diabetes. *Semin Vasc Med* 2: pp. 21–31, 2002.

[85]    Clerk LH, Rattigan S, and Clark MG. Lipid infusion impairs physiologic insulin-mediated capillary recruitment and muscle glucose uptake in vivo. *Diabetes* 51: pp. 1138–45, 2002.

[86]    Clough GF, Bennett AR, and Church MK. Effects of H1 antagonists on the cutaneous

vascular response to histamine and bradykinin: a study using scanning laser Doppler imaging. *Br J Dermatol* 138: pp. 806–14, 1998.

[87]    Cockett MI, Murphy G, Birch ML, O'Connell JP, Crabbe T, Millican AT, Hart IR, and Docherty AJ. Matrix metalloproteinases and metastatic cancer. *Biochem Soc Symp* 63: pp. 295–313, 1998.

[88]    Collard CD, Lekowski R, Jordan JE, Agah A, and Stahl GL. Complement activation following oxidative stress. *Mol Immunol* 36: pp. 941–8, 1999.

[89]    Colucci-Guyon E, Portier MM, Dunia I, Paulin D, Pournin S, and Babinet C. Mice lacking vimentin develop and reproduce without an obvious phenotype. *Cell* 79: pp. 679–94, 1994.

[90]    Cornejo CJ, Winn RK, and Harlan JM. Anti-adhesion therapy. *Adv Pharmacol* 39: pp. 99–142, 1997.

[91]    Cotran RS and Remensnyder JP. The structural basis of increased vascular permeabiligy after graded thermal injury—light and electron microscopic studies. *Ann N Y Acad Sci* 150: pp. 495–509, 1968.

[92]    Crone C and Olesen SP. Electrical resistance of brain microvascular endothelium. *Brain Res* 241: pp. 49–55, 1982.

[93]    Curreri PW, Rayfield DL, Vaught M, and Baxter CR. Extravascular fibrinogen degradation in experimental burn wounds: a source of fibrin split products. *Surgery* 77: pp. 86–91, 1975.

[94]    Curry FE and Adamson RH. Transendothelial pathways in venular microvessels exposed to agents which increase permeability: the gaps in our knowledge. *Microcirculation* 6: pp. 3–5, 1999.

[95]    Curry FE, Huxley VH, and Adamson RH. Permeability of single capillaries to intermediate-sized colored solutes. *Am J Physiol* 245: pp. H495–505, 1983.

[96]    Curry FR. Microvascular solute and water transport. *Microcirculation* 12: pp. 17–31, 2005.

[97]    Curry FR and Adamson RH. Vascular permeability modulation at the cell, microvessel, or whole organ level: towards closing gaps in our knowledge. *Cardiovasc Res,* 2010.

[98]    Curtis TM, McKeown-Longo PJ, Vincent PA, Homan SM, Wheatley EM, and Saba TM. Fibronectin attenuates increased endothelial monolayer permeability after RGD peptide, anti-alpha 5 beta 1, or TNF-alpha exposure. *Am J Physiol* 269: pp. L248–60, 1995.

[99]    Curtis TM, Rotundo RF, Vincent PA, McKeown-Longo PJ, and Saba TM. TNF-alpha-induced matrix Fn disruption and decreased endothelial integrity are independent of Fn proteolysis. *Am J Physiol* 275: pp. L126–38, 1998.

[100]   Curtis TM and Scholfield CN. The role of lipids and protein kinase Cs in the pathogenesis of diabetic retinopathy. *Diabetes Metab Res Rev* 20: pp. 28–43, 2004.

[101]   Daly C, Wong V, Burova E, Wei Y, Zabski S, Griffiths J, Lai KM, Lin HC, Ioffe E,

Yancopoulos GD, and Rudge JS. Angiopoietin-1 modulates endothelial cell function and gene expression via the transcription factor FKHR (FOXO1). *Genes Dev* 18: pp. 1060–71, 2004.

[102]   Davidson MK, Russ PK, Glick GG, Hoffman LH, Chang MS, and Haselton FR. Reduced expression of the adherens junction protein cadherin-5 in a diabetic retina. *Am J Ophthalmol* 129: pp. 267–9, 2000.

[103]   Davidson SM and Duchen MR. Endothelial mitochondria: contributing to vascular function and disease. *Circ Res* 100: pp. 1128–41, 2007.

[104]   Davis S, Aldrich TH, Jones PF, Acheson A, Compton DL, Jain V, Ryan TE, Bruno J, Radziejewski C, Maisonpierre PC, and Yancopoulos GD. Isolation of angiopoietin-1, a ligand for the TIE2 receptor, by secretion-trap expression cloning. *Cell* 87: pp. 1161–9, 1996.

[105]   Deitch EA. Multiple organ failure. Pathophysiology and potential future therapy. *Ann Surg* 216: pp. 117–34, 1992.

[106]   Dejana E and Lauri D. Biochemical and functional characteristics of integrins: a new family of adhesive receptors present in hematopoietic cells. *Haematologica* 75: pp. 1–6, 1990.

[107]   del Zoppo GJ and Milner R. Integrin-matrix interactions in the cerebral microvasculature. *Arterioscler Thromb Vasc Biol* 26: pp. 1966–1975, 2006.

[108]   Dell'Omo G, Penno G, Pucci L, Mariani M, Del Prato S, and Pedrinelli R. Abnormal capillary permeability and endothelial dysfunction in hypertension with comorbid metabolic syndrome. *Atherosclerosis* 172: pp. 383–9, 2004.

[109]   Demling RH. The burn edema process: current concepts. *J Burn Care Rehabil* 26: pp. 207–27, 2005.

[110]   Dhainaut JF, Yan SB, and Claessens YE. Protein C/activated protein C pathway: overview of clinical trial results in severe sepsis. *Crit Care Med* 32: pp. S194–201, 2004.

[111]   Dimopoulos GJ, Semba S, Kitazawa K, Eto M, and Kitazawa T. Ca2+-dependent rapid Ca2+ sensitization of contraction in arterial smooth muscle. *Circ Res* 100: pp. 121–9, 2007.

[112]   DiStasi MR and Ley K. Opening the flood-gates: how neutrophil–endothelial interactions regulate permeability. *Trends Immunol* 30: pp. 547–56, 2009.

[113]   Doolittle RF. Determining the crystal structure of fibrinogen. *J Thromb Haemost* 2: pp. 683–9, 2004.

[114]   Drake R, Gaar KA, and Taylor AE. Estimation of the filtration coefficient of pulmonary exchange vessels. *Am J Physiol* 234: pp. H266–74, 1978.

[115]   Dudek SM, Jacobson JR, Chiang ET, Birukov KG, Wang P, Zhan X, and Garcia JG. Pulmonary endothelial cell barrier enhancement by sphingosine 1-phosphate: roles for cortactin and myosin light chain kinase. *J Biol Chem* 279: pp. 24692–700, 2004.

[116]   Duffy MJ, Maguire TM, Hill A, McDermott E, and O'Higgins N. Metalloproteinases: role in breast carcinogenesis, invasion and metastasis. *Breast Cancer Res* 2: pp. 252–7, 2000.

[117] Duffy MJ, McKiernan E, O'Donovan N, and McGowan PM. The role of ADAMs in disease pathophysiology. *Clin Chim Acta* 403: pp. 31–6, 2009.

[118] Duran WN, Breslin JW, and Sanchez FA. The NO cascade, eNOS location, and microvascular permeability. *Cardiovasc Res,* 2010.

[119] Dvorak AM, Kohn S, Morgan ES, Fox P, Nagy JA, and Dvorak HF. The vesiculo-vacuolar organelle (VVO): a distinct endothelial cell structure that provides a transcellular pathway for macromolecular extravasation. *J Leukoc Biol* 59: pp. 100–15, 1996.

[120] Dvorak HF, Brown LF, Detmar M, and Dvorak AM. Vascular permeability factor/vascular endothelial growth factor, microvascular hyperpermeability, and angiogenesis. *Am J Pathol* 146: 1029–39, 1995.

[121] Eccles SA, Box GM, Court WJ, Bone EA, Thomas W, and Brown PD. Control of lymphatic and hematogenous metastasis of a rat mammary carcinoma by the matrix metalloproteinase inhibitor batimastat (BB-94). *Cancer Res* 56: pp. 2815–22, 1996.

[122] Eliceiri BP, Paul R, Schwartzberg PL, Hood JD, Leng J, and Cheresh DA. Selective requirement for Src kinases during VEGF-induced angiogenesis and vascular permeability. *Mol Cell* 4: pp. 915–24, 1999.

[123] Eliceiri BP, Puente XS, Hood JD, Stupack DG, Schlaepfer DD, Huang XZ, Sheppard D, and Cheresh DA. Src-mediated coupling of focal adhesion kinase to integrin alpha(v)beta5 in vascular endothelial growth factor signaling. *J Cell Biol* 157: pp. 149–160, 2002.

[124] Ellerbroek SM and Stack MS. Membrane associated matrix metalloproteinases in metastasis. *Bioessays* 21: pp. 940–9, 1999.

[125] Engelhardt B and Vestweber D. The Multistep Cascade of Leukocyte Extravasation. In: *Microvascular research: [biology and pathology]*, edited by Shepro D. Amsterdam: Elsevier Academic Press, 2006: pp. 303–7.

[126] Engelhardt B and Wolburg H. Mini-review: transendothelial migration of leukocytes: through the front door or around the side of the house? *Eur J Immunol* 34: pp. 2955–63, 2004.

[127] Esmon CT. Structure and functions of the endothelial cell protein C receptor. *Crit Care Med* 32: pp. S298–301, 2004.

[128] Farese RV. Insulin-sensitive phospholipid signaling systems and glucose transport. Update II. *Exp Biol Med (Maywood)* 226: pp. 283–95, 2001.

[129] Farrell DH. Pathophysiologic roles of the fibrinogen gamma chain. *Curr Opin Hematol* 11: pp. 151–5, 2004.

[130] Feng J, Ito M, Ichikawa K, Isaka N, Nishikawa M, Hartshorne DJ, and Nakano T. Inhibitory phosphorylation site for Rho-associated kinase on smooth muscle myosin phosphatase. *J Biol Chem* 274: pp. 37385–90, 1999.

[131] Fiedler U, Reiss Y, Scharpfenecker M, Grunow V, Koidl S, Thurston G, Gale NW,

Witzenrath M, Rosseau S, Suttorp N, Sobke A, Herrmann M, Preissner KT, Vajkoczy P, and Augustin HG. Angiopoietin-2 sensitizes endothelial cells to TNF-alpha and has a crucial role in the induction of inflammation. *Nat Med* 12: pp. 235–9, 2006.

[132] Fiedler U, Scharpfenecker M, Koidl S, Hegen A, Grunow V, Schmidt JM, Kriz W, Thurston G, and Augustin HG. The Tie-2 ligand angiopoietin-2 is stored in and rapidly released upon stimulation from endothelial cell Weibel-Palade bodies. *Blood* 103: pp. 4150–56, 2004.

[133] Fijalkowska I and Tuder RM. Matrix Metalloproteinases and Their Inhibitors. In: *Microvascular research: [biology and pathology]*, edited by Shepro D. Amsterdam: Elsevier Academic Press, 2006: pp. 33–40.

[134] Finigan JH. The coagulation system and pulmonary endothelial function in acute lung injury. *Microvasc Res* 77: pp. 35–8, 2009.

[135] Finigan JH, Dudek SM, Singleton PA, Chiang ET, Jacobson JR, Camp SM, Ye SQ, and Garcia JG. Activated protein C mediates novel lung endothelial barrier enhancement: role of sphingosine 1-phosphate receptor transactivation. *J Biol Chem* 280: pp. 17286–93, 2005.

[136] Fink MP. Prostaglandins and sepsis: still a fascinating topic despite almost 40 years of research. *Am J Physiol Lung Cell Mol Physiol* 281: pp. L534–6, 2001.

[137] Finkelstein EB and D'Amore PA. VEGF-A and Its Isoforms. In: *Microvascular research: [biology and pathology]*, edited by Shepro D. Amsterdam; Boston: Elsevier Academic Press, 2006: pp. 41–6.

[138] Fischer S, Wiesnet M, Renz D, and Schaper W. $H_2O_2$ induces paracellular permeability of porcine brain-derived microvascular endothelial cells by activation of the p44/42 MAP kinase pathway. *Eur J Cell Biol* 84: pp. 687–97, 2005.

[139] Flick MJ, Du X, Witte DP, Jirouskova M, Soloviev DA, Busuttil SJ, Plow EF, and Degen JL. Leukocyte engagement of fibrin(ogen) via the integrin receptor alphaMbeta2/Mac-1 is critical for host inflammatory response in vivo. *J Clin Invest* 113: pp. 1596–606, 2004.

[140] Frokjaer-Jensen J. The endothelial vesicle system in cryofixed frog mesenteric capillaries analysed by ultrathin serial sectioning. *J Electron Microsc Tech* 19: pp. 291–304, 1991.

[141] Fukumura D, Gohongi T, Kadambi A, Izumi Y, Ang J, Yun CO, Buerk DG, Huang PL, and Jain RK. Predominant role of endothelial nitric oxide synthase in vascular endothelial growth factor-induced angiogenesis and vascular permeability. *Proc Natl Acad Sci U S A* 98: pp. 2604–9, 2001.

[142] Gabbiani G and Badonnel MC. Early changes of endothelial clefts after thermal injury. *Microvasc Res* 10: pp. 65–75, 1975.

[143] Gaffney PJ. Fibrin degradation products. A review of structures found in vitro and in vivo. *Ann N Y Acad Sci* 936: pp. 594–610, 2001.

[144]  Gao X, Kouklis P, Xu N, Minshall RD, Sandoval R, Vogel SM, and Malik AB. Reversibility of increased microvessel permeability in response to VE–cadherin disassembly. *Am J Physiol Lung Cell Mol Physiol* 279: pp. L1218–25, 2000.

[145]  Garcia JG, Davis HW, and Patterson CE. Regulation of endothelial cell gap formation and barrier dysfunction: role of myosin light chain phosphorylation. *J Cell Physiol* 163: pp. 510–22, 1995.

[146]  Garcia JG, Lazar V, Gilbert-McClain LI, Gallagher PJ, and Verin AD. Myosin light chain kinase in endothelium: molecular cloning and regulation. *Am J Respir Cell Mol Biol* 16: pp. 489–94, 1997.

[147]  Garcia JG, Liu F, Verin AD, Birukova A, Dechert MA, Gerthoffer WT, Bamberg JR, and English D. Sphingosine 1-phosphate promotes endothelial cell barrier integrity by Edg-dependent cytoskeletal rearrangement. *J Clin Invest* 108: pp. 689–701, 2001.

[148]  Garcia-Avello A, Lorente JA, Cesar-Perez J, Garcia-Frade LJ, Alvarado R, Arevalo JM, Navarro JL, and Esteban A. Degree of hypercoagulability and hyperfibrinolysis is related to organ failure and prognosis after burn trauma. *Thromb Res* 89: pp. 59–64, 1998.

[149]  Garlick DG and Renkin EM. Transport of large molecules from plasma to interstitial fluid and lymph in dogs. *Am J Physiol* 219: pp. 1595–1605, 1970.

[150]  Gaudreault N, Perrin RM, Guo M, Clanton CP, Wu MH, and Yuan SY. Counter regulatory effects of PKCbetaII and PKCdelta on coronary endothelial permeability. *Arterioscler Thromb Vasc Biol* 28: pp. 1527–1533, 2008.

[151]  Gavard J, Patel V, and Gutkind JS. Angiopoietin-1 prevents VEGF-induced endothelial permeability by sequestering Src through mDia. *Dev Cell* 14: pp. 25–36, 2008.

[152]  Ge M, Ryan TJ, Lum H, and Malik AB. Fibrinogen degradation product fragment D increases endothelial monolayer permeability. *Am J Physiol* 261: pp. L283–9, 1991.

[153]  Geiser T, Atabai K, Jarreau PH, Ware LB, Pugin J, and Matthay MA. Pulmonary edema fluid from patients with acute lung injury augments in vitro alveolar epithelial repair by an IL-1beta-dependent mechanism. *Am J Respir Crit Care Med* 163: pp. 1384–8, 2001.

[154]  Geraldes P and King GL. Activation of protein kinase C isoforms and its impact on diabetic complications. *Circ Res* 106: pp. 1319–31, 2010.

[155]  Gerber HP, Kowalski J, Sherman D, Eberhard DA, and Ferrara N. Complete inhibition of rhabdomyosarcoma xenograft growth and neovascularization requires blockade of both tumor and host vascular endothelial growth factor. *Cancer Res* 60: pp. 6253–8, 2000.

[156]  Giaever I and Keese CR. Micromotion of mammalian cells measured electrically. *Proc Natl Acad Sci U S A* 88: pp. 7896–900, 1991.

[157]  Gibran NS and Heimbach DM. Current status of burn wound pathophysiology. *Clin Plast Surg* 27: pp. 11–22, 2000.

[158] Gibran NS and Heimbach DM. Mediators in thermal injury. *Semin Nephrol* 13: pp. 344–58, 1993.

[159] Gilmont RR, Dardano A, Young M, Engle JS, Adamson BS, Smith DJ, Jr., and Rees RS. Effects of glutathione depletion on oxidant-induced endothelial cell injury. *J Surg Res* 80: pp. 62–8, 1998.

[160] Godzich M, Hodnett M, Frank JA, Su G, Pespeni M, Angel A, Howard MB, Matthay MA, and Pittet JF. Activation of the stress protein response prevents the development of pulmonary edema by inhibiting VEGF cell signaling in a model of lung ischemia-reperfusion injury in rats. *FASEB J* 20: pp. 1519–21, 2006.

[161] Goeckeler ZM and Wysolmerski RB. Myosin light chain kinase-regulated endothelial cell contraction: the relationship between isometric tension, actin polymerization, and myosin phosphorylation. *J Cell Biol* 130: pp. 613–27, 1995.

[162] Goldin A, Beckman JA, Schmidt AM, and Creager MA. Advanced glycation end products: sparking the development of diabetic vascular injury. *Circulation* 114: pp. 597–605, 2006.

[163] Gomis-Ruth FX, Maskos K, Betz M, Bergner A, Huber R, Suzuki K, Yoshida N, Nagase H, Brew K, Bourenkov GP, Bartunik H, and Bode W. Mechanism of inhibition of the human matrix metalloproteinase stromelysin-1 by TIMP-1. *Nature* 389: pp. 77–81, 1997.

[164] Goodfellow J. Microvascular heart disease in diabetes mellitus. *Diabetologia* 40 Suppl 2: pp. S130–3, 1997.

[165] Gorovoy M, Neamu R, Niu J, Vogel S, Predescu D, Miyoshi J, Takai Y, Kini V, Mehta D, Malik AB, and Voyno-Yasenetskaya T. RhoGDI-1 modulation of the activity of monomeric RhoGTPase RhoA regulates endothelial barrier function in mouse lungs. *Circ Res* 101: pp. 50–8, 2007.

[166] Granger DN and Schmid-Schönbein GW. *Physiology and pathophysiology of leukocyte adhesion*. New York: Oxford University Press, 1995, p. xvi, 501 p.

[167] Gratton JP, Bernatchez P, and Sessa WC. Caveolae and caveolins in the cardiovascular system. *Circ Res* 94: pp. 1408–17, 2004.

[168] Grega GJ and Adamski SW. Differential effects of inhibitors of cellular function on inflammatory mediator-stimulated increases in vascular permeability. *Microcirc Endothelium Lymphatics* 7: pp. 217–44, 1991.

[169] Gueugniaud PY, Carsin H, Bertin-Maghit M, and Petit P. Current advances in the initial management of major thermal burns. *Intensive Care Med* 26: pp. 848–56, 2000.

[170] Guo M, Daines D, Tang J, Shen Q, Perrin RM, Takada Y, Yuan SY, and Wu MH. Fibrinogen-gamma C-terminal fragments induce endothelial barrier dysfunction and microvascular leak via integrin-mediated and RhoA-dependent mechanism. *Arterioscler Thromb Vasc Biol* 29: pp. 394–400, 2009.

[171] Guo M, Wu MH, Granger HJ, and Yuan SY. Focal adhesion kinase in neutrophil-induced microvascular hyperpermeability. *Microcirculation* 12: pp. 223–32, 2005.

[172] Guo M, Wu MH, Korompai F, and Yuan SY. Upregulation of PKC genes and isozymes in cardiovascular tissues during early stages of experimental diabetes. *Physiol Genomics* 12: pp. 139–46, 2003.

[173] Guo RF and Ward PA. Role of oxidants in lung injury during sepsis. *Antioxid Redox Signal* 9: pp. 1991–2002, 2007.

[174] Guyton AC and Lindsay AE. Effect of elevated left atrial pressure and decreased plasma protein concentration on the development of pulmonary edema. *Circ Res* 7: pp. 649–57, 1959.

[175] Hamacher J, Lucas R, Lijnen HR, Buschke S, Dunant Y, Wendel A, Grau GE, Suter PM, and Ricou B. Tumor necrosis factor-alpha and angiostatin are mediators of endothelial cytotoxicity in bronchoalveolar lavages of patients with acute respiratory distress syndrome. *Am J Respir Crit Care Med* 166: pp. 651–6, 2002.

[176] Hammel HT and Schlegel WM. Osmosis and solute-solvent drag: fluid transport and fluid exchange in animals and plants. *Cell Biochem Biophys* 42: pp. 277–345, 2005.

[177] Harburger DS and Calderwood DA. Integrin signalling at a glance. *J Cell Sci* 122: pp. 159–63, 2009.

[178] Haritoglou C, Kook D, Neubauer A, Wolf A, Priglinger S, Strauss R, Gandorfer A, Ulbig M, and Kampik A. Intravitreal bevacizumab (Avastin) therapy for persistent diffuse diabetic macular edema. *Retina* 26: pp. 999–1005, 2006.

[179] Hartshorne DJ, Ito M, and Erdodi F. Myosin light chain phosphatase: subunit composition, interactions and regulation. *J Muscle Res Cell Motil* 19: pp. 325–41, 1998.

[180] Haselton FR and Alexander JS. Platelets and a platelet-released factor enhance endothelial barrier. *Am J Physiol* 263: pp. L670–8, 1992.

[181] Haselton FR, Dworska EJ, and Hoffman LH. Glucose-induced increase in paracellular permeability and disruption of beta-receptor signaling in retinal endothelium. *Invest Ophthalmol Vis Sci* 39: pp. 1676–84, 1998.

[182] Hashizume H, Baluk P, Morikawa S, McLean JW, Thurston G, Roberge S, Jain RK, and McDonald DM. Openings between defective endothelial cells explain tumor vessel leakiness. *Am J Pathol* 156: pp. 1363–80, 2000.

[183] Hauck G and Schroer H. [Vital microscopy studies on the localization of protein permeability in the terminal stream bed of homoiotherms]. *Pflugers Arch* 312: pp. 32–44, 1969.

[184] Hawkins BT and Davis TP. The blood-brain barrier/neurovascular unit in health and disease. *Pharmacol Rev* 57: pp. 173–85, 2005.

[185] He Z and King GL. Microvascular complications of diabetes. *Endocrinol Metab Clin North Am* 33: pp. 215–38, xi–xii, 2004.

[186] Hecquet CM, Ahmmed GU, and Malik AB. TRPM2 Channel Regulates Endothelial Barrier Function. *Adv Exp Med Biol* 661: pp. 155–67, 2010.

[187] Hecquet CM, Ahmmed GU, Vogel SM, and Malik AB. Role of TRPM2 channel in mediating H2O2-induced Ca2+ entry and endothelial hyperpermeability. *Circ Res* 102: pp. 347–55, 2008.

[188] Hecquet CM and Malik AB. Role of H(2)O(2)-activated TRPM2 calcium channel in oxidant-induced endothelial injury. *Thromb Haemost* 101: pp. 619–25, 2009.

[189] Heissig B, Nishida C, Tashiro Y, Sato Y, Ishihara M, Ohki M, Gritli I, Rosenkvist J, and Hattori K. Role of neutrophil-derived matrix metalloproteinase-9 in tissue regeneration. *Histol Histopathol* 25: pp. 765–70, 2010.

[190] Hempel A, Lindschau C, Maasch C, Mahn M, Bychkov R, Noll T, Luft FC, and Haller H. Calcium antagonists ameliorate ischemia-induced endothelial cell permeability by inhibiting protein kinase C. *Circulation* 99: pp. 2523–9, 1999.

[191] Hempel A, Maasch C, Heintze U, Lindschau C, Dietz R, Luft FC, and Haller H. High glucose concentrations increase endothelial cell permeability via activation of protein kinase C alpha. *Circ Res* 81: pp. 363–71, 1997.

[192] Hermant B, Weidenhaupt M, Heyraud S, and Gulino-Debrac D. Vascular Endothelial Cadherin and Neutrophil Transmigration. In: *Microvascular research: [biology and pathology]*, edited by Shepro D. Amsterdam: Elsevier Academic Press, 2006: pp. 309–16.

[193] Hidalgo M and Eckhardt SG. Development of matrix metalloproteinase inhibitors in cancer therapy. *J Natl Cancer Inst* 93: pp. 178–93, 2001.

[194] Hirase T, Kawashima S, Wong EY, Ueyama T, Rikitake Y, Tsukita S, Yokoyama M, and Staddon JM. Regulation of tight junction permeability and occludin phosphorylation by Rhoa-p160ROCK-dependent and -independent mechanisms. *J Biol Chem* 276: pp. 10423–31, 2001.

[195] Hixenbaugh EA, Goeckeler ZM, Papaiya NN, Wysolmerski RB, Silverstein SC, and Huang AJ. Stimulated neutrophils induce myosin light chain phosphorylation and isometric tension in endothelial cells. *Am J Physiol* 273: pp. H981–8, 1997.

[196] Hobson KG, Young KM, Ciraulo A, Palmieri TL, and Greenhalgh DG. Release of abdominal compartment syndrome improves survival in patients with burn injury. *J Trauma* 53: pp. 1129–33; discussion 1133–1124, 2002.

[197] Hodivala-Dilke KM, Reynolds AR, and Reynolds LE. Integrins in angiogenesis: multitalented molecules in a balancing act. *Cell Tissue Res* 314: 131–44, 2003.

[198] Horan JT and Francis CW. Fibrin degradation products, fibrin monomer and soluble fibrin in disseminated intravascular coagulation. *Semin Thromb Hemost* 27: pp. 657–66, 2001.

[199] Hotchkiss RS and Karl IE. The pathophysiology and treatment of sepsis. *N Engl J Med* 348: pp. 138–50, 2003.

[200]  Hu C, Ahmed M, Melia TJ, Sollner TH, Mayer T, and Rothman JE. Fusion of cells by flipped SNAREs. *Science* 300: pp. 1745–9, 2003.

[201]  Hu G and Minshall RD. Regulation of transendothelial permeability by Src kinase. *Microvasc Res* 77: pp. 21–25, 2009.

[202]  Hu G, Place AT, and Minshall RD. Regulation of endothelial permeability by Src kinase signaling: vascular leakage versus transcellular transport of drugs and macromolecules. *Chem Biol Interact* 171: pp. 177–89, 2008.

[203]  Hu G, Vogel SM, Schwartz DE, Malik AB, and Minshall RD. Intercellular adhesion molecule-1-dependent neutrophil adhesion to endothelial cells induces caveolae-mediated pulmonary vascular hyperpermeability. *Circ Res* 102: pp. e120–31, 2008.

[204]  Hu X and Weinbaum S. A new view of Starling's hypothesis at the microstructural level. *Microvasc Res* 58: pp. 281–304, 1999.

[205]  Huang Q, Xu W, Ustinova E, Wu M, Childs E, Hunter F, and Yuan S. Myosin light chain kinase-dependent microvascular hyperpermeability in thermal injury. *Shock* 20: pp. 363–8, 2003.

[206]  Huang Q and Yuan Y. Interaction of PKC and NOS in signal transduction of microvascular hyperpermeability. *Am J Physiol* 273: pp. H2442–51, 1997.

[207]  Huber JD, VanGilder RL, and Houser KA. Streptozotocin-induced diabetes progressively increases blood-brain barrier permeability in specific brain regions in rats. *Am J Physiol Heart Circ Physiol* 291: pp. H2660–8, 2006.

[208]  Huxley VH, Curry FE, and Adamson RH. Quantitative fluorescence microscopy on single capillaries: alpha-lactalbumin transport. *Am J Physiol* 252: pp. H188–197, 1987.

[209]  Iba T, Gando S, Murata A, Kushimoto S, Saitoh D, Eguchi Y, Ohtomo Y, Okamoto K, Koseki K, Mayumi T, Ikeda T, Ishhikura H, Ueyama M, Ogura Y, Endo S, and Shimazaki S. Predicting the severity of systemic inflammatory response syndrome (SIRS)-associated coagulopathy with hemostatic molecular markers and vascular endothelial injury markers. *J Trauma* 63: pp. 1093–8, 2007.

[210]  Idris I, Gray S, and Donnelly R. Protein kinase C activation: isozyme-specific effects on metabolism and cardiovascular complications in diabetes. *Diabetologia* 44: pp. 659–73, 2001.

[211]  Ingber DE. Mechanical signaling and the cellular response to extracellular matrix in angiogenesis and cardiovascular physiology. *Circ Res* 91: pp. 877–87, 2002.

[212]  Jacobson JR and Garcia JG. Novel therapies for microvascular permeability in sepsis. *Curr Drug Targets* 8: pp. 509–14, 2007.

[213]  Jeon BH, Khanday F, Deshpande S, Haile A, Ozaki M, and Irani K. Tie-ing the anti-inflammatory effect of angiopoietin-1 to inhibition of NF-kappaB. *Circ Res* 92: pp. 586–8, 2003.

[214] Jezierska A and Motyl T. Matrix metalloproteinase-2 involvement in breast cancer progression: a mini-review. *Med Sci Monit* 15: pp. RA32–40, 2009.

[215] Jho D, Mehta D, Ahmmed G, Gao XP, Tiruppathi C, Broman M, and Malik AB. Angiopoietin-1 opposes VEGF-induced increase in endothelial permeability by inhibiting TRPC1-dependent Ca2 influx. *Circ Res* 96: pp. 1282–90, 2005.

[216] Johnson SA, Balboa RS, Dessel BH, Monto RW, Siegesmund KA, and Greenwalt TJ. The Mechanism of the Endothelial Supporting Function of Intact Platelets. *Exp Mol Pathol* 34: pp. 115–27, 1964.

[217] Joy SV, Scates AC, Bearelly S, Dar M, Taulien CA, Goebel JA, and Cooney MJ. Ruboxistaurin, a protein kinase C beta inhibitor, as an emerging treatment for diabetes microvascular complications. *Ann Pharmacother* 39: pp. 1693–9, 2005.

[218] Joyce DE, Gelbert L, Ciaccia A, DeHoff B, and Grinnell BW. Gene expression profile of antithrombotic protein c defines new mechanisms modulating inflammation and apoptosis. *J Biol Chem* 276: pp. 11199–203, 2001.

[219] Joyce DE, Nelson DR, and Grinnell BW. Leukocyte and endothelial cell interactions in sepsis: relevance of the protein C pathway. *Crit Care Med* 32: pp. S280–6, 2004.

[220] Ju H, Zou R, Venema VJ, and Venema RC. Direct interaction of endothelial nitric-oxide synthase and caveolin-1 inhibits synthase activity. *J Biol Chem* 272: pp. 18522–5, 1997.

[221] Kang N, Alexander G, Park JK, Maasch C, Buchwalow I, Luft FC, and Haller H. Differential expression of protein kinase C isoforms in streptozotocin-induced diabetic rats. *Kidney Int* 56: pp. 1737–50, 1999.

[222] Kargozaran H, Yuan SY, Breslin JW, Watson KD, Gaudreault N, Breen A, and Wu MH. A role for endothelial-derived matrix metalloproteinase-2 in breast cancer cell transmigration across the endothelial-basement membrane barrier. *Clin Exp Metastasis* 24: pp. 495–502, 2007.

[223] Kataoka N, Iwaki K, Hashimoto K, Mochizuki S, Ogasawara Y, Sato M, Tsujioka K, and Kajiya F. Measurements of endothelial cell-to-cell and cell-to-substrate gaps and micromechanical properties of endothelial cells during monocyte adhesion. *Proc Natl Acad Sci U S A* 99: pp. 15638–43, 2002.

[224] Keck PJ, Hauser SD, Krivi G, Sanzo K, Warren T, Feder J, and Connolly DT. Vascular permeability factor, an endothelial cell mitogen related to PDGF. *Science* 246: pp. 1309–12, 1989.

[225] Kedem O and Katchalsky A. Thermodynamic analysis of the permeability of biological membranes to non-electrolytes. *Biochim Biophys Acta* 27: pp. 229–46, 1958.

[226] Kemp-Harper B and Schmidt HH. cGMP in the vasculature. *Handb Exp Pharmacol* 447–467, 2009.

[227] Kerlin B, Cooley BC, Isermann BH, Hernandez I, Sood R, Zogg M, Hendrickson SB, Mosesson MW, Lord S, and Weiler H. Cause–effect relation between hyperfibrinogenemia and vascular disease. *Blood* 103: pp. 1728–34, 2004.

[228] Kern DF and Malik AB. Microvascular albumin permeability in isolated perfused lung: effects of EDTA. *J Appl Physiol* 58: pp. 372–5, 1985.

[229] Kevil CG, Okayama N, and Alexander JS. H(2)O(2)-mediated permeability II: importance of tyrosine phosphatase and kinase activity. *Am J Physiol Cell Physiol* 281: pp. C1940–7, 2001.

[230] Khatri JJ, Joyce KM, Brozovich FV, and Fisher SA. Role of myosin phosphatase isoforms in cGMP-mediated smooth muscle relaxation. *J Biol Chem* 276: pp. 37250–7, 2001.

[231] Kheifets V and Mochly-Rosen D. Insight into intra- and inter-molecular interactions of PKC: design of specific modulators of kinase function. *Pharmacol Res* 55: pp. 467–76, 2007.

[232] Khokha R, Voura E, and Hill RP. Tumor progression and Metastasis: Cellular, Molecular and Microenvironmental Factors. In: *The Basic Science of Oncology*, edited by Tannock I. New York: McGraw-Hill, Medical Pub. Division, 2005: pp. 205–230.

[233] Kim M, Carman CV, and Springer TA. Bidirectional transmembrane signaling by cytoplasmic domain separation in integrins. *Science* 301: pp. 1720–5, 2003.

[234] Kim MH, Curry FR, and Simon SI. Dynamics of neutrophil extravasation and vascular permeability are uncoupled during aseptic cutaneous wounding. *Am J Physiol Cell Physiol* 296: pp. C848–56, 2009.

[235] Kinashi T. Intracellular signalling controlling integrin activation in lymphocytes. *Nat Rev Immunol* 5: pp. 546–59, 2005.

[236] Kinsky MP, Guha SC, Button BM, and Kramer GC. The role of interstitial starling forces in the pathogenesis of burn edema. *J Burn Care Rehabil* 19: pp. 1–9, 1998.

[237] Kohn S, Nagy JA, Dvorak HF, and Dvorak AM. Pathways of macromolecular tracer transport across venules and small veins. Structural basis for the hyperpermeability of tumor blood vessels. *Lab Invest* 67: pp. 596–607, 1992.

[238] Kolosova IA, Ma SF, Adyshev DM, Wang P, Ohba M, Natarajan V, Garcia JG, and Verin AD. Role of CPI-17 in the regulation of endothelial cytoskeleton. *Am J Physiol Lung Cell Mol Physiol* 287: pp. L970–80, 2004.

[239] Komarova Y and Malik AB. Regulation of endothelial permeability via paracellular and transcellular transport pathways. *Annu Rev Physiol* 72: pp. 463–93, 2010.

[240] Kontos CD, Cha EH, York JD, and Peters KG. The endothelial receptor tyrosine kinase Tie1 activates phosphatidylinositol 3-kinase and Akt to inhibit apoptosis. *Mol Cell Biol* 22: pp. 1704–13, 2002.

[241] Kornerup K, Nordestgaard BG, Feldt-Rasmussen B, Borch-Johnsen K, Jensen KS, and Jensen JS. Transvascular low-density lipoprotein transport in patients with diabetes mellitus (type 2): a noninvasive in vivo isotope technique. *Arterioscler Thromb Vasc Biol* 22: pp. 1168–74, 2002.

[242] Koval M and Battacharya J. Vascular Gap Junctions. In: *Microvascular research: [biology and pathology]*, edited by Shepro D. Amsterdam: Elsevier Academic Press, 2006: pp. 317–20.

[243] Koya D and King GL. Protein kinase C activation and the development of diabetic complications. *Diabetes* 47: pp. 859–66, 1998.

[244] Kubes P. Nitric oxide affects microvascular permeability in the intact and inflamed vasculature. *Microcirculation* 2: pp. 235–44, 1995.

[245] Kubes P. The complexities of leukocyte recruitment. *Semin Immunol* 14: pp. 65–72, 2002.

[246] Kubes P and Granger DN. Nitric oxide modulates microvascular permeability. *Am J Physiol* 262: pp. H611–15, 1992.

[247] Kumar P, Shen Q, Pivetti CD, Lee ES, Wu MH, and Yuan SY. Molecular mechanisms of endothelial hyperpermeability: implications in inflammation. *Expert Rev Mol Med* 11: p. e19, 2009.

[248] Lafleur MA, Handsley MM, and Edwards DR. Metalloproteinases and their inhibitors in angiogenesis. *Expert Rev Mol Med* 5: pp. 1–39, 2003.

[249] Laine GA and Granger HJ. Microvascular, interstitial, and lymphatic interactions in normal heart. *Am J Physiol* 249: pp. H834–42, 1985.

[250] Lampugnani MG, Resnati M, Dejana E, and Marchisio PC. The role of integrins in the maintenance of endothelial monolayer integrity. *J Cell Biol* 112: pp. 479–90, 1991.

[251] Landis EM. Micro-injection studies of capillary permeability: II. The relation between capillary pressure and the rate at which fluid passes through the walls of single capillaries. *Am J Physiol* 82: pp. 217–38, 1927.

[252] Landis EM, Jonas L, Angevine M, and Erb W. The passage of fluid and protein through the human capillary wall during venous congestion. *J Clin Invest* 11: pp. 717–34, 1932.

[253] Laszlo F, Whittle BJ, and Moncada S. Interactions of constitutive nitric oxide with PAF and thromboxane on rat intestinal vascular integrity in acute endotoxaemia. *Br J Pharmacol* 113: pp. 1131–6, 1994.

[254] Laurens N, Koolwijk P, and de Maat MP. Fibrin structure and wound healing. *J Thromb Haemost* 4: pp. 932–9, 2006.

[255] Lawrence MB and Springer TA. Leukocytes roll on a selectin at physiologic flow rates: distinction from and prerequisite for adhesion through integrins. *Cell* 65: pp. 859–73, 1991.

[256] Lazar V and Garcia JG. A single human myosin light chain kinase gene (MLCK; MYLK). *Genomics* 57: pp. 256–67, 1999.

[257] Leco KJ, Khokha R, Pavloff N, Hawkes SP, and Edwards DR. Tissue inhibitor of metallo-proteinases-3 (TIMP-3) is an extracellular matrix-associated protein with a distinctive pattern of expression in mouse cells and tissues. *J Biol Chem* 269: pp. 9352–60, 1994.

[258] Lee ES, Shen Q, Pitts RL, Guo M, Wu MH, and Yuan SY. Vein tissue expression of matrix metalloproteinase as biomarker for hemodialysis arteriovenous fistula maturation. *Vasc Endovascular Surg* 44: pp. 674–9, 2010.

[259] Lee FY, Covello KL, Castaneda S, Hawken DR, Kan D, Lewin A, Wen ML, Ryseck RP, Fairchild CR, Fargnoli J, and Kramer R. Synergistic antitumor activity of ixabepilone (BMS-247550) plus bevacizumab in multiple in vivo tumor models. *Clin Cancer Res* 14: pp. 8123–31, 2008.

[260] Lee H, Volonte D, Galbiati F, Iyengar P, Lublin DM, Bregman DB, Wilson MT, Campos-Gonzalez R, Bouzahzah B, Pestell RG, Scherer PE, and Lisanti MP. Constitutive and growth factor-regulated phosphorylation of caveolin-1 occurs at the same site (Tyr-14) in vivo: identification of a c-Src/Cav-1/Grb7 signaling cassette. *Mol Endocrinol* 14: pp. 1750–75, 2000.

[261] Leung DW, Cachianes G, Kuang WJ, Goeddel DV, and Ferrara N. Vascular endothelial growth factor is a secreted angiogenic mitogen. *Science* 246: pp. 1306–9, 1989.

[262] Levick JR. *An introduction to cardiovascular physiology*. London: Hodder Arnold, 2010, p. xii, 414 p.

[263] Levick JR and Michel CC. A densitometric method for determining the filtration coefficients of single capillaries in the frog mesentery. *Microvasc Res* 13: pp. 141–51, 1977.

[264] Levick JR and Michel CC. Microvascular fluid exchange and the revised Starling principle. *Cardiovasc Res* 87: pp. 198–210, 2010.

[265] Levy MN, Pappano AJ, and Berne RM. *Cardiovascular physiology*. Philadelphia, PA: Mosby Elsevier, 2007, p. xiv, 269 p.

[266] Lewis RE and Granger HJ. Diapedesis and the permeability of venous microvessels to protein macromolecules: the impact of leukotriene B4 (LTB4). *Microvasc Res* 35: pp. 27–47, 1988.

[267] Lewis RE and Granger HJ. Neutrophil-dependent mediation of microvascular permeability. *Fed Proc* 45: pp. 109–13, 1986.

[268] Ley K. Adhesion Molecules and the Recruitment of Leukocytes in Postcapillary Venules. In: *Microvascular research: [biology and pathology]*, edited by Shepro D. Amsterdam: Elsevier Academic Press, 2006: pp. 321–6.

[269] Ley K and Arfors KE. Segmental differences of microvascular permeability for FITC-dextrans measured in the hamster cheek pouch. *Microvasc Res* 31: pp. 84–99, 1986.

[270] Li JM and Shah AM. Endothelial cell superoxide generation: regulation and relevance for

cardiovascular pathophysiology. *Am J Physiol Regul Integr Comp Physiol* 287: pp. R1014–30, 2004.

[271]   Li S, Seitz R, and Lisanti MP. Phosphorylation of caveolin by src tyrosine kinases. The alpha-isoform of caveolin is selectively phosphorylated by v-Src in vivo. *J Biol Chem* 271: pp. 3863–8, 1996.

[272]   Li X, Stankovic M, Bonder CS, Hahn CN, Parsons M, Pitson SM, Xia P, Proia RL, Vadas MA, and Gamble JR. Basal and angiopoietin-1-mediated endothelial permeability is regulated by sphingosine kinase-1. *Blood* 111: pp. 3489–97, 2008.

[273]   Liaw PC. Endogenous protein C activation in patients with severe sepsis. *Crit Care Med* 32: pp. S214–8, 2004.

[274]   Lin MI and Sessa WC. Roles of nitric oxide in the microcirculation. In: *Microvascular research: [biology and pathology]*, edited by Shepro D. Amsterdam: Elsevier Academic Press, 2006: pp. 229–32.

[275]   Lin MI, Yu J, Murata T, and Sessa WC. Caveolin-1-deficient mice have increased tumor microvascular permeability, angiogenesis, and growth. *Cancer Res* 67: pp. 2849–56, 2007.

[276]   Lincoln TM. Myosin phosphatase regulatory pathways: different functions or redundant functions? *Circ Res* 100: pp. 10–12, 2007.

[277]   Liu SM, Magnusson KE, and Sundqvist T. Microtubules are involved in transport of macromolecules by vesicles in cultured bovine aortic endothelial cells. *J Cell Physiol* 156: pp. 311–6, 1993.

[278]   Liu X, Wang J, Takeda N, Binaglia L, Panagia V, and Dhalla NS. Changes in cardiac protein kinase C activities and isozymes in streptozotocin-induced diabetes. *Am J Physiol* 277: pp. E798–804, 1999.

[279]   Lo SK, Burhop KE, Kaplan JE, and Malik AB. Role of platelets in maintenance of pulmonary vascular permeability to protein. *Am J Physiol* 254: pp. H763–71, 1988.

[280]   London NR, Whitehead KJ, and Li DY. Endogenous endothelial cell signaling systems maintain vascular stability. *Angiogenesis* 12: pp. 149–58, 2009.

[281]   Lord ST. Fibrinogen and fibrin: scaffold proteins in hemostasis. *Curr Opin Hematol* 14: pp. 236–41, 2007.

[282]   Lowell CA and Soriano P. Knockouts of Src-family kinases: stiff bones, wimpy T cells, and bad memories. *Genes Dev* 10: pp. 1845–57, 1996.

[283]   Lucas N and Day ML. The role of the disintegrin metalloproteinase ADAM15 in prostate cancer progression. *J Cell Biochem* 106: pp. 967–74, 2009.

[284]   Lund T. The 1999 Everett Idris Evans memorial lecture. Edema generation following thermal injury: an update. *J Burn Care Rehabil* 20: pp. 445–52, 1999.

[285]   Lund T, Onarheim H, and Reed RK. Pathogenesis of edema formation in burn injuries. *World J Surg* 16: pp. 2–9, 1992.

[286]  Luo BH, Carman CV, and Springer TA. Structural basis of integrin regulation and signaling. *Annu Rev Immunol* 25: pp. 619–47, 2007.

[287]  Luscinskas FW and Lawler J. Integrins as dynamic regulators of vascular function. *FASEB J* 8: pp. 929–38, 1994.

[288]  Luscinskas FW, Ma S, Nusrat A, Parkos CA, and Shaw SK. Leukocyte transendothelial migration: a junctional affair. *Semin Immunol* 14: pp. 105–13, 2002.

[289]  Lush CW and Kvietys PR. Microvascular dysfunction in sepsis. *Microcirculation* 7: pp. 83–101, 2000.

[290]  Maniatis NA, Brovkovych V, Allen SE, John TA, Shajahan AN, Tiruppathi C, Vogel SM, Skidgel RA, Malik AB, and Minshall RD. Novel mechanism of endothelial nitric oxide synthase activation mediated by caveolae internalization in endothelial cells. *Circ Res* 99: pp. 870–7, 2006.

[291]  Manning G, Whyte DB, Martinez R, Hunter T, and Sudarsanam S. The protein kinase complement of the human genome. *Science* 298: pp. 1912–34, 2002.

[292]  Manon-Jensen T, Itoh Y, and Couchman JR. Proteoglycans in health and disease: the multiple roles of syndecan shedding. *FEBS J* 277: pp. 3876–89, 2010.

[293]  Maretzky T, Yang G, Ouerfelli O, Overall CM, Worpenberg S, Hassiepen U, Eder J, and Blobel CP. Characterization of the catalytic activity of the membrane-anchored metalloproteinase ADAM15 in cell-based assays. *Biochem J* 420: pp. 105–13, 2009.

[294]  Martin J, Eynstone LV, Davies M, Williams JD, and Steadman R. The role of ADAM 15 in glomerular mesangial cell migration. *J Biol Chem* 277: pp. 33683–9, 2002.

[295]  Masson V, de la Ballina LR, Munaut C, Wielockx B, Jost M, Maillard C, Blacher S, Bajou K, Itoh T, Itohara S, Werb Z, Libert C, Foidart JM, and Noel A. Contribution of host MMP-2 and MMP-9 to promote tumor vascularization and invasion of malignant keratinocytes. *FASEB J* 19: pp. 234–36, 2005.

[296]  Matsumoto T, Wada H, Nobori T, Nakatani K, Onishi K, Nishikawa M, Shiku H, Kazahaya Y, Sawai T, Koike K, and Matsuda M. Elevated plasma levels of fibrin degradation products by granulocyte-derived elastase in patients with disseminated intravascular coagulation. *Clin Appl Thromb Hemost* 11: pp. 391–400, 2005.

[297]  Mayhan WG and Joyner WL. The effect of altering the external calcium concentration and a calcium channel blocker, verapamil, on microvascular leaky sites and dextran clearance in the hamster cheek pouch. *Microvasc Res* 28: pp. 159–79, 1984.

[298]  McDonagh PF and Hokama JY. Microvascular perfusion and transport in the diabetic heart. *Microcirculation* 7: pp. 163–81, 2000.

[299]  McGillem LL and Garcia JG. Role of the Endothelial Cell Cytoskeleton in Microvascular Function. In: *Microvascular research : [biology and pathology]*, edited by Shepro D. Amsterdam: Elsevier Academic Press, 2006: pp. 247–54.

[300] McLaughlin JN, Shen L, Holinstat M, Brooks JD, Dibenedetto E, and Hamm HE. Functional selectivity of G protein signaling by agonist peptides and thrombin for the protease-activated receptor-1. *J Biol Chem* 280: pp. 25048–59, 2005.

[301] Mehta D and Malik AB. Signaling mechanisms regulating endothelial permeability. *Physiol Rev* 86: pp. 279–367, 2006.

[302] Mehta D, Rahman A, and Malik AB. Protein kinase C-alpha signals rho-guanine nucleotide dissociation inhibitor phosphorylation and rho activation and regulates the endothelial cell barrier function. *J Biol Chem* 276: pp. 22614–20, 2001.

[303] Meyn MA, 3rd and Smithgall TE. Small molecule inhibitors of Lck: the search for specificity within a kinase family. *Mini Rev Med Chem* 8: pp. 628–37, 2008.

[304] Michel CC. Starling: the formulation of his hypothesis of microvascular fluid exchange and its significance after 100 years. *Exp Physiol* 82: pp. 1–30, 1997.

[305] Michel CC and Curry FE. Microvascular permeability. *Physiol Rev* 79: pp. 703–61, 1999.

[306] Michel CC, Mason JC, Curry FE, Tooke JE, and Hunter PJ. A development of the Landis technique for measuring the filtration coefficient of individual capillaries in the frog mesentery. *Q J Exp Physiol Cogn Med Sci* 59: pp. 283–309, 1974.

[307] Middleton J, Neil S, Wintle J, Clark-Lewis I, Moore H, Lam C, Auer M, Hub E, and Rot A. Transcytosis and surface presentation of IL-8 by venular endothelial cells. *Cell* 91: pp. 385–95, 1997.

[308] Miles AA and Miles EM. Vascular reactions to histamine, histamine-liberator and leukotaxine in the skin of guinea-pigs. *J Physiol* 118: pp. 228–57, 1952.

[309] Minnear FL. Platelet phospholipids tighten the vascular endothelial barrier. In: *Microvascular research: [biology and pathology]*, edited by Shepro D. Amsterdam: Elsevier Academic Press, 2006: pp. 345–51.

[310] Minnear FL, Patil S, Bell D, Gainor JP, and Morton CA. Platelet lipid(s) bound to albumin increases endothelial electrical resistance: mimicked by LPA. *Am J Physiol Lung Cell Mol Physiol* 281: pp. L1337–44, 2001.

[311] Minshall RD, Sessa WC, Stan RV, Anderson RG, and Malik AB. Caveolin regulation of endothelial function. *Am J Physiol Lung Cell Mol Physiol* 285: pp. L1179–83, 2003.

[312] Minshall RD, Tiruppathi C, Vogel SM, Niles WD, Gilchrist A, Hamm HE, and Malik AB. Endothelial cell-surface gp60 activates vesicle formation and trafficking via G(i)-coupled Src kinase signaling pathway. *J Cell Biol* 150: pp. 1057–70, 2000.

[313] Moitra J, Evenoski C, Sammani S, Wadgaonkar R, Turner JR, Ma SF, and Garcia JG. A transgenic mouse with vascular endothelial over-expression of the non-muscle myosin light chain kinase-2 isoform is susceptible to inflammatory lung injury: role of sexual dimorphism and age. *Transl Res* 151: pp. 141–53, 2008.

[314]  Moncada S and Higgs EA. Nitric oxide and the vascular endothelium. *Handb Exp Pharmacol*, pp. 213–54, 2006.

[315]  Monti M, Donnini S, Giachetti A, Mochly-Rosen D, and Ziche M. deltaPKC inhibition or varepsilonPKC activation repairs endothelial vascular dysfunction by regulating eNOS post-translational modification. *J Mol Cell Cardiol* 48: pp. 746–56, 2010.

[316]  Morikawa S, Baluk P, Kaidoh T, Haskell A, Jain RK, and McDonald DM. Abnormalities in pericytes on blood vessels and endothelial sprouts in tumors. *Am J Pathol* 160: pp. 985–1000, 2002.

[317]  Moser M and Leo O. Key concepts in immunology. *Vaccine* 28 Suppl 3: pp. C2–13, 2010.

[318]  Mosesson MW. Fibrinogen and fibrin structure and functions. *J Thromb Haemost* 3: pp. 1894–1904, 2005.

[319]  Mosesson MW. Fibrinogen gamma chain functions. *J Thromb Haemost* 1: pp. 231–8, 2003.

[320]  Moy AB, Sheldon R, Lindsley K, Shasby S, and Shasby DM. Centripetal tension and endothelial retraction. *Chest* 105: pp. 107S–8S, 1994.

[321]  Muranyi A, Derkach D, Erdodi F, Kiss A, Ito M, and Hartshorne DJ. Phosphorylation of Thr695 and Thr850 on the myosin phosphatase target subunit: inhibitory effects and occurrence in A7r5 cells. *FEBS Lett* 579: pp. 6611–15, 2005.

[322]  Murphy G, Houbrechts A, Cockett MI, Williamson RA, O'Shea M, and Docherty AJ. The N-terminal domain of tissue inhibitor of metalloproteinases retains metalloproteinase inhibitory activity. *Biochemistry* 30: pp. 8097–102, 1991.

[323]  Murphy KD, Lee JO, and Herndon DN. Current pharmacotherapy for the treatment of severe burns. *Expert Opin Pharmacother* 4: pp. 369–84, 2003.

[324]  Nagase H, Visse R, and Murphy G. Structure and function of matrix metalloproteinases and TIMPs. *Cardiovasc Res* 69: pp. 562–73, 2006.

[325]  Naglich JG, Jure-Kunkel M, Gupta E, Fargnoli J, Henderson AJ, Lewin AC, Talbott R, Baxter A, Bird J, Savopoulos R, Wills R, Kramer RA, and Trail PA. Inhibition of angiogenesis and metastasis in two murine models by the matrix metalloproteinase inhibitor, BMS-275291. *Cancer Res* 61: pp. 8480–5, 2001.

[326]  Najy AJ, Day KC, and Day ML. ADAM15 supports prostate cancer metastasis by modulating tumor cell–endothelial cell interaction. *Cancer Res* 68: pp. 1092–9, 2008.

[327]  Najy AJ, Day KC, and Day ML. The ectodomain shedding of E-cadherin by ADAM15 supports ErbB receptor activation. *J Biol Chem* 283: pp. 18393–401, 2008.

[328]  Nakamura Y and Wayland H. Macromolecular transport in the cat mesentery. *Microvasc Res* 9: pp. 1–21, 1975.

[329]  Nath D, Slocombe PM, Stephens PE, Warn A, Hutchinson GR, Yamada KM, Docherty

AJ, and Murphy G. Interaction of metargidin (ADAM-15) with alphavbeta3 and alpha-5beta1 integrins on different haemopoietic cells. *J Cell Sci* 112 (Pt 4): pp. 579–87, 1999.

[330]   Nathan C. Neutrophils and immunity: challenges and opportunities. *Nat Rev Immunol* 6: pp. 173–82, 2006.

[331]   Ng ES and Kubes P. The physiology of *S*-nitrosothiols: carrier molecules for nitric oxide. *Can J Physiol Pharmacol* 81: pp. 759–64, 2003.

[332]   Nobe K, Miyatake M, Sone T, and Honda K. High-glucose-altered endothelial cell function involves both disruption of cell-to-cell connection and enhancement of force development. *J Pharmacol Exp Ther* 318: pp. 530–39, 2006.

[333]   Nosal'ova V, Drabikova K, Jancinova V, Pecivova J, Macickova T, Petrikova M, Sotnikova R, and Nosal R. Effect of H1 antihistamines in a model of mesenteric ischaemia/reperfusion. *Inflamm Res* 57 Suppl 1: pp. S55–6, 2008.

[334]   Nwariaku FE, Liu Z, Zhu X, Turnage RH, Sarosi GA, and Terada LS. Tyrosine phosphorylation of vascular endothelial cadherin and the regulation of microvascular permeability. *Surgery* 132: pp. 180–5, 2002.

[335]   Okutani D, Lodyga M, Han B, and Liu M. Src protein tyrosine kinase family and acute inflammatory responses. *Am J Physiol Lung Cell Mol Physiol* 291: pp. L129–41, 2006.

[336]   Palade GE, Simionescu M, and Simionescu N. Structural aspects of the permeability of the microvascular endothelium. *Acta Physiol Scand Suppl* 463: pp. 11–32, 1979.

[337]   Pappenheimer JR. Passage of molecules through capillary wals. *Physiol Rev* 33: pp. 387–423, 1953.

[338]   Pappenheimer JR, Renkin EM, and Borrero LM. Filtration, diffusion and molecular sieving through peripheral capillary membranes; a contribution to the pore theory of capillary permeability. *Am J Physiol* 167: pp. 13–46, 1951.

[339]   Pappenheimer JR and Soto-Rivera A. Effective osmotic pressure of the plasma proteins and other quantities associated with the capillary circulation in the hindlimbs of cats and dogs. *Am J Physiol* 152: pp. 471–91, 1948.

[340]   Pardridge WM. *Introduction to the blood-brain barrier: methodology, biology and pathology.* Cambridge, NY: Cambridge University Press, 1998, p. xiv, 486 p.

[341]   Parikh SM, Mammoto T, Schultz A, Yuan HT, Christiani D, Karumanchi SA, and Sukhatme VP. Excess circulating angiopoietin-2 may contribute to pulmonary vascular leak in sepsis in humans. *PLoS Med* 3: p. e46, 2006.

[342]   Parsons JT. Focal adhesion kinase: the first ten years. *J Cell Sci* 116: pp. 1409–16, 2003.

[343]   Parsons ME and Ganellin CR. Histamine and its receptors. *Br J Pharmacol* 147 Suppl 1: pp. S127–35, 2006.

[344]   Patterson CE, Lum H, Schaphorst KL, Verin AD, and Garcia JG. Regulation of endothe-

lial barrier function by the cAMP-dependent protein kinase. *Endothelium* 7: pp. 287–308, 2000.

[345]   Patterson CE, Rhoades RA, and Garcia JG. Evans blue dye as a marker of albumin clearance in cultured endothelial monolayer and isolated lung. *J Appl Physiol* 72: pp. 865–73, 1992.

[346]   Pavlaki M and Zucker S. Matrix metalloproteinase inhibitors (MMPIs): the beginning of phase I or the termination of phase III clinical trials. *Cancer Metastasis Rev* 22: pp. 177–203, 2003.

[347]   Pearse DB, Brower RG, Adkinson NF, Jr., and Sylvester JT. Spontaneous injury in isolated sheep lungs: role of perfusate leukocytes and platelets. *J Appl Physiol* 66: pp. 1287–96, 1989.

[348]   Petit V and Thiery JP. Focal adhesions: structure and dynamics. *Biol Cell* 92: pp. 477–94, 2000.

[349]   Pfarr KM, Debrah AY, Specht S, and Hoerauf A. Filariasis and lymphoedema. *Parasite Immunol* 31: pp. 664–72, 2009.

[350]   Piedra J, Miravet S, Castano J, Palmer HG, Heisterkamp N, Garcia de Herreros A, and Dunach M. p120 Catenin-associated Fer and Fyn tyrosine kinases regulate beta-catenin Tyr-142 phosphorylation and beta-catenin-alpha-catenin interaction. *Mol Cell Biol* 23: pp. 2287–97, 2003.

[351]   Pierce KL, Premont RT, and Lefkowitz RJ. Seven-transmembrane receptors. *Nat Rev Mol Cell Biol* 3: pp. 639–50, 2002.

[352]   Pitt RM, Parker JC, Jurkovich GJ, Taylor AE, and Curreri PW. Analysis of altered capillary pressure and permeability after thermal injury. *J Surg Res* 42: pp. 693–702, 1987.

[353]   Piulats J and Mitjans F. Angiogenesis switch pathways. In: *Principles of molecular oncology*, edited by Bronchud MH. Totowa, NJ: Humana Press, 2008: pp. 239–56.

[354]   Pizurki L, Zhou Z, Glynos K, Roussos C, and Papapetropoulos A. Angiopoietin-1 inhibits endothelial permeability, neutrophil adherence and IL-8 production. *Br J Pharmacol* 139: pp. 329–36, 2003.

[355]   Pocock TM, Foster RR, and Bates DO. Evidence of a role for TRPC channels in VEGF-mediated increased vascular permeability in vivo. *Am J Physiol Heart Circ Physiol* 286: pp. H1015–26, 2004.

[356]   Polverini PJ, Cotran PS, Gimbrone MA, Jr., and Unanue ER. Activated macrophages induce vascular proliferation. *Nature* 269: pp. 804–6, 1977.

[357]   Prager MD, Baxter CR, and Hartline B. Proteolytic activity in burn wound exudates and comparison of fibrin degradation products and protease inhibitors in exudates and sera. *J Burn Care Rehabil* 15: pp. 130–6, 1994.

[358] Prasain N and Stevens T. The actin cytoskeleton in endothelial cell phenotypes. *Microvasc Res* 77: pp. 53–63, 2009.

[359] Predescu D, Horvat R, Predescu S, and Palade GE. Transcytosis in the continuous endothelium of the myocardial microvasculature is inhibited by N-ethylmaleimide. *Proc Natl Acad Sci U S A* 91: pp. 3014–18, 1994.

[360] Predescu D and Palade GE. Plasmalemmal vesicles represent the large pore system of continuous microvascular endothelium. *Am J Physiol* 265: pp. H725–33, 1993.

[361] Predescu SA, Predescu DN, and Malik AB. Molecular determinants of endothelial transcytosis and their role in endothelial permeability. *Am J Physiol Lung Cell Mol Physiol* 293: pp. L823–42, 2007.

[362] Predescu SA, Predescu DN, and Palade GE. Endothelial transcytotic machinery involves supramolecular protein-lipid complexes. *Mol Biol Cell* 12: pp. 1019–33, 2001.

[363] Predescu SA, Predescu DN, Timblin BK, Stan RV, and Malik AB. Intersectin regulates fission and internalization of caveolae in endothelial cells. *Mol Biol Cell* 14: pp. 4997–5010, 2003.

[364] Pun PB, Lu J, and Moochhala S. Involvement of ROS in BBB dysfunction. *Free Radic Res* 43: pp. 348–64, 2009.

[365] Purchio AF, Erikson E, Brugge JS, and Erikson RL. Identification of a polypeptide encoded by the avian sarcoma virus src gene. *Proc Natl Acad Sci U S A* 75: pp. 1567–71, 1978.

[366] Pyne S and Pyne N. Sphingosine 1-phosphate signalling via the endothelial differentiation gene family of G-protein-coupled receptors. *Pharmacol Ther* 88: pp. 115–31, 2000.

[367] Qian J, Zhang Q, Church JE, Stepp DW, Rudic RD, and Fulton DJ. Role of local production of endothelium-derived nitric oxide on cGMP signaling and *S*-nitrosylation. *Am J Physiol Heart Circ Physiol* 298: pp. H112–8, 2010.

[368] Qiao RL, Yan W, Lum H, and Malik AB. Arg-Gly-Asp peptide increases endothelial hydraulic conductivity: comparison with thrombin response. *Am J Physiol* 269: pp. C110–7, 1995.

[369] Quadri SK, Bhattacharjee M, Parthasarathi K, Tanita T, and Bhattacharya J. Endothelial barrier strengthening by activation of focal adhesion kinase. *J Biol Chem* 278: pp. 13342–9, 2003.

[370] Ramirez MM, Kim DD, and Duran WN. Protein kinase C modulates microvascular permeability through nitric oxide synthase. *Am J Physiol* 271: pp. H1702–5, 1996.

[371] Rask-Madsen C and King GL. Proatherosclerotic mechanisms involving protein kinase C in diabetes and insulin resistance. *Arterioscler Thromb Vasc Biol* 25: pp. 487–96, 2005.

[372] Razani B, Engelman JA, Wang XB, Schubert W, Zhang XL, Marks CB, Macaluso F, Russell RG, Li M, Pestell RG, Di Vizio D, Hou H, Jr., Kneitz B, Lagaud G, Christ GJ,

Edelmann W, and Lisanti MP. Caveolin-1 null mice are viable but show evidence of hyper-proliferative and vascular abnormalities. *J Biol Chem* 276: pp. 38121–38, 2001.

[373] Rebres RA, Cho E, Rotundo RF, and Saba TM. Reduced in vivo plasma fibronectin content of lung matrix during postoperative sepsis. *Am J Physiol* 271: pp. L409–18, 1996.

[374] Reed RK and Rubin K. Transcapillary exchange: role and importance of the interstitial fluid pressure and the extracellular matrix. *Cardiovasc Res,* 2010.

[375] Reiss K, Maretzky T, Ludwig A, Tousseyn T, de Strooper B, Hartmann D, and Saftig P. ADAM10 cleavage of N-cadherin and regulation of cell–cell adhesion and beta-catenin nuclear signalling. *EMBO J* 24: pp. 742–52, 2005.

[376] Reitsma S, Slaaf DW, Vink H, van Zandvoort MA, and oude Egbrink MG. The endothelial glycocalyx: composition, functions, and visualization. *Pflugers Arch* 454: pp. 345–59, 2007.

[377] Renkin EM. Cellular aspects of transvascular exchange: a 40-year perspective. *Microcirculation* 1: pp. 157–67, 1994.

[378] Renkin EM and Curry FE. Transport of water and solutes across capillary endothelium. In: *Membrane transport in biology*, edited by Giebisch GH, Tosteson DC, and Ussing HH. Berlin: Springer-Verlag, 1978, p. v.

[379] Reutershan J and Ley K. Bench-to-bedside review: acute respiratory distress syndrome—how neutrophils migrate into the lung. *Crit Care* 8: pp. 453–61, 2004.

[380] Reynolds AB and Roczniak-Ferguson A. Emerging roles for p120-catenin in cell adhesion and cancer. *Oncogene* 23: pp. 7947–56, 2004.

[381] Reynoso R, Perrin RM, Breslin JW, Daines DA, Watson KD, Watterson DM, Wu MH, and Yuan S. A role for long chain myosin light chain kinase (MLCK-210) in microvascular hyperpermeability during severe burns. *Shock* 28: pp. 589–95, 2007.

[382] Robinson GS, Ju M, Shih SC, Xu X, McMahon G, Caldwell RB, and Smith LE. Nonvascular role for VEGF: VEGFR-1, 2 activity is critical for neural retinal development. *FASEB J* 15: pp. 1215–7, 2001.

[383] Rodrigues SF and Granger DN. Role of blood cells in ischemia-reperfusion induced endothelial barrier failure. *Cardiovasc Res* 87: pp. 291–9, 2010.

[384] Rosenberg GA and Yang Y. Vasogenic edema due to tight junction disruption by matrix metalloproteinases in cerebral ischemia. *Neurosurg Focus* 22: p. E4, 2007.

[385] Rowland FN, Donovan MJ, Picciano PT, Wilner GD, and Kreutzer DL. Fibrin-mediated vascular injury. Identification of fibrin peptides that mediate endothelial cell retraction. *Am J Pathol* 117: pp. 418–28, 1984.

[386] Rudini N and Dejana E. Adherens junctions. *Curr Biol* 18: pp. R1080–2, 2008.

[387] Sahin U, Weskamp G, Kelly K, Zhou HM, Higashiyama S, Peschon J, Hartmann D, Saftig

P, and Blobel CP. Distinct roles for ADAM10 and ADAM17 in ectodomain shedding of six EGFR ligands. *J Cell Biol* 164: pp. 769–79, 2004.

[388]   Salameh A, Zinn M, and Dhein S. High D-glucose induces alterations of endothelial cell structure in a cell-culture model. *J Cardiovasc Pharmacol* 30: pp. 182–90, 1997.

[389]   Sallee JL, Wittchen ES, and Burridge K. Regulation of cell adhesion by protein-tyrosine phosphatases: II. Cell–cell adhesion. *J Biol Chem* 281: pp. 16189–92, 2006.

[390]   Sammani S, Moreno-Vinasco L, Mirzapoiazova T, Singleton PA, Chiang ET, Evenoski CL, Wang T, Mathew B, Husain A, Moitra J, Sun X, Nunez L, Jacobson JR, Dudek SM, Natarajan V, and Garcia JG. Differential Effects of S1P Receptors on Airway and Vascular Barrier Function in the Murine Lung. *Am J Respir Cell Mol Biol*, 2009.

[391]   Sampathkumar R, Balasubramanyam M, Rema M, Premanand C, and Mohan V. A novel advanced glycation index and its association with diabetes and microangiopathy. *Metabolism* 54: pp. 1002–7, 2005.

[392]   Sanchez FA, Kim DD, Duran RG, Meininger CJ, and Duran WN. Internalization of eNOS via caveolae regulates PAF-induced inflammatory hyperpermeability to macromolecules. *Am J Physiol Heart Circ Physiol* 295: pp. H1642–8, 2008.

[393]   Sanchez FA, Rana R, Kim DD, Iwahashi T, Zheng R, Lal BK, Gordon DM, Meininger CJ, and Duran WN. Internalization of eNOS and NO delivery to subcellular targets determine agonist-induced hyperpermeability. *Proc Natl Acad Sci U S A* 106: pp. 6849–53, 2009.

[394]   Sanchez FA, Savalia NB, Duran RG, Lal BK, Boric MP, and Duran WN. Functional significance of differential eNOS translocation. *Am J Physiol Heart Circ Physiol* 291: pp. H1058–64, 2006.

[395]   Sandoval R, Malik AB, Minshall RD, Kouklis P, Ellis CA, and Tiruppathi C. Ca(2+) signalling and PKCalpha activate increased endothelial permeability by disassembly of VE–cadherin junctions. *J Physiol* 533: pp. 433–45, 2001.

[396]   Sans E, Delachanal E, and Duperray A. Analysis of the roles of ICAM-1 in neutrophil transmigration using a reconstituted mammalian cell expression model: implication of ICAM-1 cytoplasmic domain and Rho-dependent signaling pathway. *J Immunol* 166: pp. 544–51, 2001.

[397]   Sargiacomo M, Scherer PE, Tang Z, Kubler E, Song KS, Sanders MC, and Lisanti MP. Oligomeric structure of caveolin: implications for caveolae membrane organization. *Proc Natl Acad Sci U S A* 92: pp. 9407–11, 1995.

[398]   Sayeed MM. Neutrophil signaling alteration: an adverse inflammatory response after burn shock. *Medicina (B Aires)* 58: pp. 386–92, 1998.

[399]   Scheppke L, Aguilar E, Gariano RF, Jacobson R, Hood J, Doukas J, Cao J, Noronha G, Yee

S, Weis S, Martin MB, Soll R, Cheresh DA, and Friedlander M. Retinal vascular permeability suppression by topical application of a novel VEGFR2/Src kinase inhibitor in mice and rabbits. *J Clin Invest* 118: pp. 2337–46, 2008.

[400]  Schmidt AM, Yan SD, Yan SF, and Stern DM. The biology of the receptor for advanced glycation end products and its ligands. *Biochim Biophys Acta* 1498: pp. 99–111, 2000.

[401]  Schulz B, Pruessmeyer J, Maretzky T, Ludwig A, Blobel CP, Saftig P, and Reiss K. ADAM10 regulates endothelial permeability and T-cell transmigration by proteolysis of vascular endothelial cadherin. *Circ Res* 102: pp. 1192–201, 2008.

[402]  Schulze C, Smales C, Rubin LL, and Staddon JM. Lysophosphatidic acid increases tight junction permeability in cultured brain endothelial cells. *J Neurochem* 68: pp. 991–1000, 1997.

[403]  Scott JA and King GL. Oxidative stress and antioxidant treatment in diabetes. *Ann N Y Acad Sci* 1031: pp. 204–13, 2004.

[404]  Seiki M. The cell surface: the stage for matrix metalloproteinase regulation of migration. *Curr Opin Cell Biol* 14: pp. 624–32, 2002.

[405]  Selkurt EE. *Physiology*. Boston: Little, Brown, 1984, p. ix, 691 p.

[406]  Senger DR, Galli SJ, Dvorak AM, Perruzzi CA, Harvey VS, and Dvorak HF. Tumor cells secrete a vascular permeability factor that promotes accumulation of ascites fluid. *Science* 219: pp. 983–5, 1983.

[407]  Sessa WC. The nitric oxide synthase family of proteins. *J Vasc Res* 31: 131–43, 1994.

[408]  Shajahan AN, Timblin BK, Sandoval R, Tiruppathi C, Malik AB, and Minshall RD. Role of Src-induced dynamin-2 phosphorylation in caveolae-mediated endocytosis in endothelial cells. *J Biol Chem* 279: pp. 20392–400, 2004.

[409]  Shasby DM, Ries DR, Shasby SS, and Winter MC. Histamine stimulates phosphorylation of adherens junction proteins and alters their link to vimentin. *Am J Physiol Lung Cell Mol Physiol* 282: pp. L1330–8, 2002.

[410]  Shea SM, Caulfield JB, and Burke JF. Microvascular ultrastructure in thermal injury: a reconsideration of the role of mediators. *Microvasc Res* 5: pp. 87–96, 1973.

[411]  Sheldon R, Moy A, Lindsley K, Shasby S, and Shasby DM. Role of myosin light-chain phosphorylation in endothelial cell retraction. *Am J Physiol* 265: pp. L606–12, 1993.

[412]  Shen Q, Lee ES, Pitts RL, Wu MH, and Yuan SY. Tissue inhibitor of metalloproteinase-2 regulates matrix metalloproteinase-2-mediated endothelial barrier dysfunction and breast cancer cell transmigration through lung microvascular endothelial cells. *Mol Cancer Res* 8: pp. 939–51, 2010.

[413]  Shen Q, Rigor RR, Pivetti CD, Wu MH, and Yuan SY. Myosin light chain kinase in microvascular endothelial barrier function. *Cardiovasc Res* 87: pp. 272–80, 2010.

[414]  Shen Q, Wu MH, and Yuan SY. Endothelial contractile cytoskeleton and microvascular permeability. *Cell Health and Cytoskeleton* 1: pp. 43–50, 2009.

[415]  Sherwood L. *Human physiology : from cells to systems.* Australia; United States: Brooks/Cole, Cengage Learning, 2010, p. 1 v. (various pagings).

[416]  Shikata Y, Birukov KG, Birukova AA, Verin A, and Garcia JG. Involvement of site-specific FAK phosphorylation in sphingosine-1 phosphate- and thrombin-induced focal adhesion remodeling: role of Src and GIT. *FASEB J* 17: pp. 2240–9, 2003.

[417]  Shimaoka M, Xiao T, Liu JH, Yang Y, Dong Y, Jun CD, McCormack A, Zhang R, Joachimiak A, Takagi J, Wang JH, and Springer TA. Structures of the alpha L I domain and its complex with ICAM-1 reveal a shape-shifting pathway for integrin regulation. *Cell* 112: pp. 99–111, 2003.

[418]  Shivanna M, Rajashekhar G, and Srinivas SP. Barrier dysfunction of the corneal endothelium in response to TNF-alpha: role of p38 MAP kinase. *Invest Ophthalmol Vis Sci* 51: pp. 1575–82, 2010.

[419]  Shivanna M and Srinivas SP. Elevated cAMP opposes (TNF-alpha)-induced loss in the barrier integrity of corneal endothelium. *Mol Vis* 16: pp. 1781–90, 2010.

[420]  Shivanna M and Srinivas SP. Microtubule stabilization opposes the (TNF-alpha)-induced loss in the barrier integrity of corneal endothelium. *Exp Eye Res* 89: pp. 950–9, 2009.

[421]  Shore AC. The microvasculature in type 1 diabetes. *Semin Vasc Med* 2: pp. 9–20, 2002.

[422]  Siflinger-Birnboim A, Del Vecchio PJ, Cooper JA, Blumenstock FA, Shepard JM, and Malik AB. Molecular sieving characteristics of the cultured endothelial monolayer. *J Cell Physiol* 132: pp. 111–7, 1987.

[423]  Simionescu N, Simionescu M, and Palade GE. Open junctions in the endothelium of the postcapillary venules of the diaphragm. *J Cell Biol* 79: pp. 27–44, 1978.

[424]  Simons K and Ikonen E. Functional rafts in cell membranes. *Nature* 387: pp. 569–72, 1997.

[425]  Simpson-Haidaris PJ, and Rybarczyk B. Tumors and fibrinogen. The role of fibrinogen as an extracellular matrix protein. *Ann N Y Acad Sci* 936: pp. 406–25, 2001.

[426]  Sledge GW, Jr., Qulali M, Goulet R, Bone EA, and Fife R. Effect of matrix metalloproteinase inhibitor batimastat on breast cancer regrowth and metastasis in athymic mice. *J Natl Cancer Inst* 87: pp. 1546–50, 1995.

[427]  Spindler V, Schlegel N, and Waschke J. Role of GTPases in control of microvascular permeability. *Cardiovasc Res,* 2010.

[428]  Starling EH. On the absorption of fluids from the connective tissue spaces. *J Physiol* 19: pp. 312–26, 1896.

[429]  Stasek JE, Jr., Patterson CE, and Garcia JG. Protein kinase C phosphorylates caldesmon77

and vimentin and enhances albumin permeability across cultured bovine pulmonary artery endothelial cell monolayers. *J Cell Physiol* 153: pp. 62–75, 1992.

[430]   Stenina OI. Regulation of vascular genes by glucose. *Curr Pharm Des* 11: pp. 2367–81, 2005.

[431]   Stocker W, Gomis-Ruth FX, Bode W, and Zwilling R. Implications of the three-dimensional structure of astacin for the structure and function of the astacin family of zinc-endopeptidases. *Eur J Biochem* 214: 215–231, 1993.

[432]   Stokes KY, Cooper D, Tailor A, and Granger DN. Hypercholesterolemia promotes inflammation and microvascular dysfunction: role of nitric oxide and superoxide. *Free Radic Biol Med* 33: 1026–1036, 2002.

[433]   Strongin AY, Marmer BL, Grant GA, and Goldberg GI. Plasma membrane-dependent activation of the 72-kDa type IV collagenase is prevented by complex formation with TIMP-2. *J Biol Chem* 268: 14033–14039, 1993.

[434]   Sullivan SR, Ahmadi AJ, Singh CN, Sires BS, Engrav LH, Gibran NS, Heimbach DM, and Klein MB. Elevated orbital pressure: another untoward effect of massive resuscitation after burn injury. *J Trauma* 60: 72–76, 2006.

[435]   Sun C, Wu MH, Guo M, Day ML, Lee ES, and Yuan SY. ADAM15 regulates endothelial permeability and neutrophil migration via Src/ERK1/2 signalling. *Cardiovasc Res* 2010.

[436]   Sun H, Breslin JW, Zhu J, Yuan SY, and Wu MH. Rho and ROCK signaling in VEGF-induced microvascular endothelial hyperpermeability. *Microcirculation* 13: pp. 237–47, 2006.

[437]   Suratt BT and Parsons PE. Mechanisms of acute lung injury/acute respiratory distress syndrome. *Clin Chest Med* 27: pp. 579–89; abstract viii, 2006.

[438]   Tarbell JM. Shear stress and the endothelial transport barrier. *Cardiovasc Res,* 2010.

[439]   Taylor AE. Capillary fluid filtration. Starling forces and lymph flow. *Circ Res* 49: pp. 557–75, 1981.

[440]   Thomas SM and Brugge JS. Cellular functions regulated by Src family kinases. *Annu Rev Cell Dev Biol* 13: pp. 513–609, 1997.

[441]   Tilton RG. Diabetic vascular dysfunction: links to glucose-induced reductive stress and VEGF. *Microsc Res Tech* 57: pp. 390–407, 2002.

[442]   Tinsley JH, Teasdale NR, and Yuan SY. Involvement of PKCdelta and PKD in pulmonary microvascular endothelial cell hyperpermeability. *Am J Physiol Cell Physiol* 286: pp. C105–11, 2004.

[443]   Tinsley JH, Ustinova EE, Xu W, and Yuan SY. Src-dependent, neutrophil-mediated vascular hyperpermeability and beta-catenin modification. *Am J Physiol Cell Physiol* 283: pp. C1745–51, 2002.

[444] Tiruppathi C, Ahmmed GU, Vogel SM, and Malik AB. Ca2+ signaling, TRP channels, and endothelial permeability. *Microcirculation* 13: pp. 693–708, 2006.

[445] Tiruppathi C, Malik AB, Del Vecchio PJ, Keese CR, and Giaever I. Electrical method for detection of endothelial cell shape change in real time: assessment of endothelial barrier function. *Proc Natl Acad Sci U S A* 89: pp. 7919–23, 1992.

[446] Tomlinson DR. Mitogen-activated protein kinases as glucose transducers for diabetic complications. *Diabetologia* 42: pp. 1271–81, 1999.

[447] Tooke JE. Microvascular function in human diabetes. A physiological perspective. *Diabetes* 44: pp. 721–6, 1995.

[448] Trache A, Trzeciakowski JP, Gardiner L, Sun Z, Muthuchamy M, Guo M, Yuan SY, and Meininger GA. Histamine effects on endothelial cell fibronectin interaction studied by atomic force microscopy. *Biophys J* 89: pp. 2888–98, 2005.

[449] Tuttle KR, Bakris GL, Toto RD, McGill JB, Hu K, and Anderson PW. The effect of ruboxistaurin on nephropathy in type 2 diabetes. *Diabetes Care* 28: pp. 2686–90, 2005.

[450] Tyagi N, Roberts AM, Dean WL, Tyagi SC, and Lominadze D. Fibrinogen induces endothelial cell permeability. *Mol Cell Biochem* 307: pp. 13–22, 2008.

[451] Ugarova TP, Lishko VK, Podolnikova NP, Okumura N, Merkulov SM, Yakubenko VP, Yee VC, Lord ST, and Haas TA. Sequence gamma 377-395(P2), but not gamma 190-202(P1), is the binding site for the alpha MI-domain of integrin alpha M beta 2 in the gamma C-domain of fibrinogen. *Biochemistry* 42: pp. 9365–73, 2003.

[452] Unterberg A, Wahl M, and Baethmann A. Effects of bradykinin on permeability and diameter of pial vessels in vivo. *J Cereb Blood Flow Metab* 4: pp. 574–85, 1984.

[453] Valensi P, Cohen-Boulakia F, Attali JR, and Behar A. Changes in capillary permeability in diabetic patients. *Clin Hemorheol Microcirc* 17: pp. 389–94, 1997.

[454] Van de Wouwer M and Conway EM. Novel functions of thrombomodulin in inflammation. *Crit Care Med* 32: pp. S254–61, 2004.

[455] van der Heijden M, Pickkers P, van Nieuw Amerongen GP, van Hinsbergh VW, Bouw MP, van der Hoeven JG, and Groeneveld AB. Circulating angiopoietin-2 levels in the course of septic shock: relation with fluid balance, pulmonary dysfunction and mortality. *Intensive Care Med* 35: pp. 1567–74, 2009.

[456] van der Heijden M, van Nieuw Amerongen GP, van Hinsbergh VW, and Groeneveld AB. The interaction of soluble Tie2 with angiopoietins and pulmonary vascular permeability in septic and nonseptic critically ill patients. *Shock* 33: pp. 263–8, 2010.

[457] van Nieuw Amerongen GP, Natarajan K, Yin G, Hoefen RJ, Osawa M, Haendeler J, Ridley AJ, Fujiwara K, van Hinsbergh VW, and Berk BC. GIT1 mediates thrombin signaling in endothelial cells: role in turnover of RhoA-type focal adhesions. *Circ Res* 94: pp. 1041–9, 2004.

[458]  van Nieuw Amerongen GP, and van Hinsbergh VW. Targets for pharmacological interven-
tion of endothelial hyperpermeability and barrier function. *Vascul Pharmacol* 39: pp. 257–72,
2002.

[459]  van Nieuw Amerongen GP, Vermeer MA, and van Hinsbergh VW. Role of RhoA and Rho
kinase in lysophosphatidic acid-induced endothelial barrier dysfunction. *Arterioscler Thromb
Vasc Biol* 20: pp. E127–33, 2000.

[460]  Vandenbroucke E, Mehta D, Minshall R, and Malik AB. Regulation of endothelial junctional permeability. *Ann N Y Acad Sci* 1123: pp. 134–45, 2008.

[461]  VanTeeffelen JW, Brands J, and Vink H. Agonist-induced impairment of glycocalyx exclu-
sion properties: contribution to coronary effects of adenosine. *Cardiovasc Res,* 2010.

[462]  Velasco G, Armstrong C, Morrice N, Frame S, and Cohen P. Phosphorylation of the regulatory subunit of smooth muscle protein phosphatase 1M at Thr850 induces its dissociation
from myosin. *FEBS Lett* 527: pp. 101–4, 2002.

[463]  Verin AD, Lazar V, Torry RJ, Labarrere CA, Patterson CE, and Garcia JG. Expression of
a novel high molecular-weight myosin light chain kinase in endothelium. *Am J Respir Cell
Mol Biol* 19: pp. 758–66, 1998.

[464]  Verin AD, Patterson CE, Day MA, and Garcia JG. Regulation of endothelial cell gap formation and barrier function by myosin-associated phosphatase activities. *Am J Physiol* 269:
pp. L99–108, 1995.

[465]  Vincent PA, Xiao K, Buckley KM, and Kowalczyk AP. VE–cadherin: adhesion at arm's
length. *Am J Physiol Cell Physiol* 286: pp. C987–97, 2004.

[466]  Vink H and Duling BR. Capillary endothelial surface layer selectively reduces plasma solute distribution volume. *Am J Physiol Heart Circ Physiol* 278: H285–289, 2000.

[467]  Vinores SA, Derevjanik NL, Mahlow J, Berkowitz BA, and Wilson CA. Electron microscopic evidence for the mechanism of blood-retinal barrier breakdown in diabetic rabbits: comparison with magnetic resonance imaging. *Pathol Res Pract* 194: pp. 497–505,
1998.

[468]  Visse R and Nagase H. Matrix metalloproteinases and tissue inhibitors of metalloproteinases: structure, function, and biochemistry. *Circ Res* 92: pp. 827–39, 2003.

[469]  Vogel SM, Minshall RD, Pilipovic M, Tiruppathi C, and Malik AB. Albumin uptake and
transcytosis in endothelial cells in vivo induced by albumin-binding protein. *Am J Physiol
Lung Cell Mol Physiol* 281: pp. L1512–22, 2001.

[470]  Wahl WL, Brandt MM, Ahrns K, Corpron CA, and Franklin GA. The utility of D-dimer
levels in screening for thromboembolic complications in burn patients. *J Burn Care Rehabil*
23: pp. 439–43, 2002.

[471]  Wainwright MS, Rossi J, Schavocky J, Crawford S, Steinhorn D, Velentza AV, Zasadzki
M, Shirinsky V, Jia Y, Haiech J, Van Eldik LJ, and Watterson DM. Protein kinase involved

in lung injury susceptibility: evidence from enzyme isoform genetic knockout and in vivo inhibitor treatment. *Proc Natl Acad Sci U S A* 100: pp. 6233–8, 2003.

[472]   Wakai A, Gleeson A, and Winter D. Role of fibrin D-dimer testing in emergency medicine. *Emerg Med J* 20: pp. 319–25, 2003.

[473]   Wang L, and Dudek SM. Regulation of vascular permeability by sphingosine 1-phosphate. *Microvasc Res* 77: pp. 39–45, 2009.

[474]   Wang N and Stamenovic D. Mechanics of vimentin intermediate filaments. *J Muscle Res Cell Motil* 23: pp. 535–40, 2002.

[475]   Wang Q, Sun AY, Simonyi A, Kalogeris TJ, Miller DK, Sun GY, and Korthuis RJ. Ethanol preconditioning protects against ischemia/reperfusion-induced brain damage: role of NADPH oxidase-derived ROS. *Free Radic Biol Med* 43: pp. 1048–60, 2007.

[476]   Ward PA and Till GO. Pathophysiologic events related to thermal injury of skin. *J Trauma* 30: pp. S75–9, 1990.

[477]   Ware LB and Matthay MA. The acute respiratory distress syndrome. *N Engl J Med* 342: pp. 1334–49, 2000.

[478]   Watterson DM, Schavocky JP, Guo L, Weiss C, Chlenski A, Shirinsky VP, Van Eldik LJ, and Haiech J. Analysis of the kinase-related protein gene found at human chromosome 3q21 in a multi-gene cluster: organization, expression, alternative splicing, and polymorphic marker. *J Cell Biochem* 75: pp. 481–91, 1999.

[479]   Wautier JL, Zoukourian C, Chappey O, Wautier MP, Guillausseau PJ, Cao R, Hori O, Stern D, and Schmidt AM. Receptor-mediated endothelial cell dysfunction in diabetic vasculopathy. Soluble receptor for advanced glycation end products blocks hyperpermeability in diabetic rats. *J Clin Invest* 97: pp. 238–43, 1996.

[480]   Way KJ, Katai N, and King GL. Protein kinase C and the development of diabetic vascular complications. *Diabet Med* 18: pp. 945–59, 2001.

[481]   Weis SM and Cheresh DA. Pathophysiological consequences of VEGF-induced vascular permeability. *Nature* 437: pp. 497–504, 2005.

[482]   Werthmann RC, von Hayn K, Nikolaev VO, Lohse MJ, and Bunemann M. Real-time monitoring of cAMP levels in living endothelial cells: thrombin transiently inhibits adenylyl cyclase 6. *J Physiol* 587: pp. 4091–104, 2009.

[483]   Wheatley EM, McKeown-Longo PJ, Vincent PA, and Saba TM. Incorporation of fibronectin into matrix decreases TNF-induced increase in endothelial monolayer permeability. *Am J Physiol* 265: pp. L148–57, 1993.

[484]   Whittaker M, Floyd CD, Brown P, and Gearing AJ. Design and therapeutic application of matrix metalloproteinase inhibitors. *Chem Rev* 99: pp. 2735–76, 1999.

[485]   Whittles CE, Pocock TM, Wedge SR, Kendrew J, Hennequin LF, Harper SJ, and Bates

DO. ZM323881, a novel inhibitor of vascular endothelial growth factor-receptor-2 tyrosine kinase activity. *Microcirculation* 9: pp. 513–22, 2002.

[486] Windsor AC, Walsh CJ, Mullen PG, Cook DJ, Fisher BJ, Blocher CR, Leeper-Woodford SK, Sugerman HJ, and Fowler AA, 3rd. Tumor necrosis factor-alpha blockade prevents neutrophil CD18 receptor upregulation and attenuates acute lung injury in porcine sepsis without inhibition of neutrophil oxygen radical generation. *J Clin Invest* 91: pp. 1459–68, 1993.

[487] Witte S, Goldenberg DM, and Schricker KT. The propagation of fluorescent dyes in the hamster cheek pouch. *Z Gesamte Exp Med* 148: pp. 72–80, 1968.

[488] Wolf BA, Williamson JR, Easom RA, Chang K, Sherman WR, and Turk J. Diacylglycerol accumulation and microvascular abnormalities induced by elevated glucose levels. *J Clin Invest* 87: pp. 31–8, 1991.

[489] Wolf MB and Watson PD. Measurement of osmotic reflection coefficient for small molecules in cat hindlimbs. *Am J Physiol* 256: pp. H282–90, 1989.

[490] Wu C, Ivars F, Anderson P, Hallmann R, Vestweber D, Nilsson P, Robenek H, Tryggvason K, Song J, Korpos E, Loser K, Beissert S, Georges-Labouesse E, and Sorokin LM. Endothelial basement membrane laminin alpha5 selectively inhibits T lymphocyte extravasation into the brain. *Nat Med* 15: pp. 519–27, 2009.

[491] Wu HM, Huang Q, Yuan Y, and Granger HJ. VEGF induces NO-dependent hyperpermeability in coronary venules. *Am J Physiol* 271: pp. H2735–9, 1996.

[492] Wu HM, Yuan Y, Zawieja DC, Tinsley J, and Granger HJ. Role of phospholipase C, protein kinase C, and calcium in VEGF-induced venular hyperpermeability. *Am J Physiol* 276: pp. H535–42, 1999.

[493] Wu MH. Endothelial focal adhesions and barrier function. *J Physiol* 569: pp. 359–66, 2005.

[494] Wu MH, Guo M, Yuan SY, and Granger HJ. Focal adhesion kinase mediates porcine venular hyperpermeability elicited by vascular endothelial growth factor. *J Physiol* 552: pp. 691–9, 2003.

[495] Wu MH, Ustinova E, and Granger HJ. Integrin binding to fibronectin and vitronectin maintains the barrier function of isolated porcine coronary venules. *J Physiol* 532: pp. 785–91, 2001.

[496] Wu MH and Yuan SY. Protein Kinases and microvascular permeability. In: *Microvascular research: [biology and pathology]*, edited by Shepro D. Amsterdam: Elsevier Academic Press, 2006: pp. 287–93.

[497] Wu MH, Yuan SY, and Granger HJ. The protein kinase MEK1/2 mediate vascular endothelial growth factor- and histamine-induced hyperpermeability in porcine coronary venules. *J Physiol* 563: pp. 95–104, 2005.

[498]  Xu X, Wang Y, Chen Z, Sternlicht MD, Hidalgo M, and Steffensen B. Matrix metalloproteinase-2 contributes to cancer cell migration on collagen. *Cancer Res* 65: pp. 130–6, 2005.

[499]  Xu Y and Yu Q. Angiopoietin-1, unlike angiopoietin-2, is incorporated into the extracellular matrix via its linker peptide region. *J Biol Chem* 276: pp. 34990–8, 2001.

[500]  Yamaji T, Fukuhara T, and Kinoshita M. Increased capillary permeability to albumin in diabetic rat myocardium. *Circ Res* 72: pp. 947–57, 1993.

[501]  Yang Z, Pandi L, and Doolittle RF. The crystal structure of fragment double-D from cross-linked lamprey fibrin reveals isopeptide linkages across an unexpected D–D interface. *Biochemistry* 41: pp. 15610–7, 2002.

[502]  Yee VC, Pratt KP, Cote HC, Trong IL, Chung DW, Davie EW, Stenkamp RE, and Teller DC. Crystal structure of a 30 kDa C-terminal fragment from the gamma chain of human fibrinogen. *Structure* 5: pp. 125–38, 1997.

[503]  Yokoyama K, Erickson HP, Ikeda Y, and Takada Y. Identification of amino acid sequences in fibrinogen gamma -chain and tenascin C C-terminal domains critical for binding to integrin alpha vbeta 3. *J Biol Chem* 275: pp. 16891–8, 2000.

[504]  Yokoyama K, Zhang XP, Medved L, and Takada Y. Specific binding of integrin alpha v beta 3 to the fibrinogen gamma and alpha E chain C-terminal domains. *Biochemistry* 38: pp. 5872–7, 1999.

[505]  Youn YK, Lalonde C, and Demling R. Oxidants and the pathophysiology of burn and smoke inhalation injury. *Free Radic Biol Med* 12: pp. 409–15, 1992.

[506]  Yu PK, Yu DY, Cringle SJ, and Su EN. Endothelial F-actin cytoskeleton in the retinal vasculature of normal and diabetic rats. *Curr Eye Res* 30: pp. 279–90, 2005.

[507]  Yuan F, Chen Y, Dellian M, Safabakhsh N, Ferrara N, and Jain RK. Time-dependent vascular regression and permeability changes in established human tumor xenografts induced by an anti-vascular endothelial growth factor/vascular permeability factor antibody. *Proc Natl Acad Sci U S A* 93: pp. 14765–70, 1996.

[508]  Yuan SY. New insights into eNOS signaling in microvascular permeability. *Am J Physiol Heart Circ Physiol* 291: pp. H1029–31, 2006.

[509]  Yuan SY. Protein kinase signaling in the modulation of microvascular permeability. *Vascul Pharmacol* 39: pp. 213–23, 2003.

[510]  Yuan SY. Protein kinase signaling in the modulation of microvascular permeability. *Vascul Pharmacol* 39: pp. 213–23, 2002.

[511]  Yuan SY. Signal transduction pathways in enhanced microvascular permeability. *Microcirculation* 7: pp. 395–403, 2000.

[512] Yuan SY, Ustinova EE, Wu MH, Tinsley JH, Xu W, Korompai FL, and Taulman AC. Protein kinase C activation contributes to microvascular barrier dysfunction in the heart at early stages of diabetes. *Circ Res* 87: pp. 412–17, 2000.

[513] Yuan Y, Chilian WM, Granger HJ, and Zawieja DC. Permeability to albumin in isolated coronary venules. *Am J Physiol* 265: pp. H543–52, 1993.

[514] Yuan Y, Granger HJ, Zawieja DC, and Chilian WM. Flow modulates coronary venular permeability by a nitric oxide-related mechanism. *Am J Physiol* 263: pp. H641–6, 1992.

[515] Yuan Y, Granger HJ, Zawieja DC, DeFily DV, and Chilian WM. Histamine increases venular permeability via a phospholipase C-NO synthase-guanylate cyclase cascade. *Am J Physiol* 264: pp. H1734–9, 1993.

[516] Yuan Y, Huang Q, and Wu HM. Myosin light chain phosphorylation: modulation of basal and agonist-stimulated venular permeability. *Am J Physiol* 272: pp. H1437–43, 1997.

[517] Yuan Y, Meng FY, Huang Q, Hawker J, and Wu HM. Tyrosine phosphorylation of paxillin/pp125FAK and microvascular endothelial barrier function. *Am J Physiol* 275: pp. H84–93, 1998.

[518] Yuan Y, Mier RA, Chilian WM, Zawieja DC, and Granger HJ. Interaction of neutrophils and endothelium in isolated coronary venules and arterioles. *Am J Physiol* 268: pp. H490–8, 1995.

[519] Zacharowski K, Zacharowski P, Reingruber S, and Petzelbauer P. Fibrin(ogen) and its fragments in the pathophysiology and treatment of myocardial infarction. *J Mol Med* 84: pp. 469–77, 2006.

[520] Zhao J, Singleton PA, Brown ME, Dudek SM, and Garcia JG. Phosphotyrosine protein dynamics in cell membrane rafts of sphingosine-1-phosphate-stimulated human endothelium: role in barrier enhancement. *Cell Signal* 21: pp. 1945–60, 2009.

[521] Zlotnik A and Yoshie O. Chemokines: a new classification system and their role in immunity. *Immunity* 12: pp. 121–7, 2000.

[522] Zweifach BW and Intaglietta M. Mechanics of fluid movement across single capillaries in the rabbit. *Microvasc Res* 1: pp. 83–101, 1968.